Computational Risk Management

Risks exist in every aspect of our lives and risk management has always been a vital topic. Most computational techniques and tools have been used for optimizing risk management and the risk management tools benefit from computational approaches. Computational intelligence models such as neural networks and support vector machine have been widely used for early warning of company bankruptcy and credit risk rating, operational research approaches such as VaR (value at risk) optimization has been standardized in managing markets and credit risk, agent-based theories are employed in supply chain risk management and various simulation techniques are employed by researchers working on problems of environmental risk management and disaster risk management. Investigation of computational tools in risk management is beneficial to both practitioners and researchers. The Computational Risk Management series is a high-quality research book series with an emphasis on computational aspects of risk management and analysis. In this series research monographs, conference proceedings are published.

More information about this series at http://www.springer.com/series/8827

David L. Olson · Desheng Wu

Predictive Data Mining Models

 Springer

David L. Olson
College of Business
University of Nebraska
Lincoln, NE
USA

Desheng Wu
School of Economics and Management
University of Chinese Academy of Sciences
Beijing
China

ISSN 2191-1436   ISSN 2191-1444   (electronic)
Computational Risk Management
ISBN 978-981-10-9645-7   ISBN 978-981-10-2543-3   (eBook)
DOI 10.1007/978-981-10-2543-3

Printed on acid-free paper

This Springer imprint is published by Springer Nature
The registered company is Springer Nature Singapore Pte Ltd.
The registered company address is: 152 Beach Road, #22-06/08 Gateway East, Singapore 189721, Singapore

Preface

Knowledge management involves application of human knowledge (epistemology) with the technological advances of our current society (computer systems) and big data, both in terms of collecting data and in analyzing it. We see three types of analytic tools. **Descriptive** analytics focus on reports of what has happened. **Predictive** analytics extend statistical and/or artificial intelligence to provide forecasting capability. It also includes classification modeling. **Diagnostic** analytics can apply analysis to sensor input to direct control systems automatically. **Prescriptive** analytics applies quantitative models to optimize systems, or at least to identify improved systems. Data mining includes descriptive and predictive modeling. Operations research includes all four. This book focuses on the forecasting component of predictive modeling, with the classification portion of prescriptive analytics demonstrated.

Book Concept

The book seeks to provide simple explanations and demonstration of some predictive tools. Chapter 3 covers simple moving average, because it is a component of ARIMA models covered later, and linear regression. This is extended in Chap. 4 to causal regression. Chapter 5 covers regression trees. Chapter 6 describes and demonstrates autoregressive integrated moving-average (ARIMA) and generalized autoregressive conditional heteroscedasticity (GARCH) models. Chapter 7 provides a quick overview of classification modeling. Chapter 8 reviews how these different kinds of predictive models provide value in the era of big data.

Models are demonstrated using business related data to include stock indices, crude oil price, and the price of gold. The style of the book is intended to be descriptive, seeking to explain how methods work, with some citations, but without deep scholarly reference. The data sets and software are all selected for widespread availability and access by any reader with computer links.

Lincoln, NE, USA   David L. Olson
Beijing, China   Desheng Wu

Acknowledgment

This work is supported by the Ministry of Science and Technology of China under Grant 2016YFC0503606, by National Natural Science Foundation of China (NSFC) grant [grant nos 71471055; 91546102], and by Chinese Academy of Sciences Frontier Scientific Research Key Project under Grant No. QYZDB-SSW-SYS021.

Contents

About the Authors

David L. Olson is the James and H.K. Stuart Chancellor's Distinguished Chair and Full Professor at the University of Nebraska. He has published research in over 150 refereed journal articles, primarily on the topic of multiple objective decision-making, information technology, supply chain risk management, and data mining. He teaches in the management information systems, management science, and operations management areas. He has authored over 20 books. He is a member of the Decision Sciences Institute, the Institute for Operations Research and Management Sciences, and the Multiple Criteria Decision Making Society. He was a Lowry Mays endowed Professor at Texas A&M University from 1999 to 2001. He was named the Raymond E. Miles Distinguished Scholar award for 2002, and was a James C. and Rhonda Seacrest Fellow from 2005 to 2006. He was named Best Enterprise Information Systems Educator by IFIP in 2006. He is a Fellow of the Decision Sciences Institute.

Desheng Wu is the distinguished professor at University of Chinese Academy of Sciences. His research interests focus on enterprise risk management, performance evaluation, and decision support system. He has published more than 80 journal papers appeared in such journals as Production and Operations Management, IEEE Transactions on Systems, Man, and Cybernetics: Systems, Risk Analysis, Decision Sciences, Decision Support Systems, European Journal of Operational Research, IEEE Transactions on Knowledge and Data Engineering, etc. He has coauthored 3 books with David L. Olson. He has served as editor/guest editors for several journals such as IEEE Transactions on Systems, Man, and Cybernetics: Part B, Omega, Computers and OR, International Journal of Production Research.

Chapter 1
Knowledge Management

Knowledge management is an overarching term referring to the ability to identify, store, and retrieve knowledge. **Identification** requires gathering the information needed and to analyze available data to make effective decisions regarding whatever the organization does. This include research, digging through records, or gathering data from wherever it can be found. **Storage** and **retrieval** of data involves database management, using many tools developed by computer science. Thus knowledge management involves understanding what knowledge is important to the organization, understanding systems important to organizational decision making, database management, and analytic tools of data mining.

The era of big data is here [1]. Davenport defines big data as:

- Data too big to fit on a single server;
- Too unstructured to fit in a row-and-column database;
- Flowing too continuously to fit into a static data warehouse;
- Having the characteristic of lacking structure.

Knowledge management (KM) needs to cope with big data by identifying and managing knowledge assets within organizations. KM is process oriented, thinking in terms of how knowledge can be acquired, as well as tools to aid decision making. Rothberg and Erickson [2] give a framework defining data as **observation**, which when put into context becomes **information**, which when processed by human understanding becomes **knowledge**. The point of big data is to analyze, converting data into insights, innovation, and business value. It can add value by providing real-time measures of performance, provide more timely analyses based on more complete data, and lead to sounder decisions [3].

We live in an environment driven by data. Amazon prospers by understanding what their customers want, and delivering content in effective ways. Wal-Mart has been very successful in using electronic means to gather sales in real time, storing 65 weeks of data in massive data warehouse systems that they intensively mine to support inventory, pricing, transportation, and other decisions related to their business. Data management also is found in governmental operations. The National

© Springer Science+Business Media Singapore 2017
D.L. Olson and D. Wu, *Predictive Data Mining Models*,
Computational Risk Management, DOI 10.1007/978-981-10-2543-3_1

Weather Service has collected unbelievable quantities of information related to weather, harnessing high-end computing power to improve weather prediction. NASA has developed a knowledge base of physical relationships enabling space activities. Waller and Fawcett [4] describe big data in terms of **volume, velocity**, and **variety**.

With respect to volume, retail sales data aggregates a firm's many cash register transactions in real time; such information can be aggregated by consumer to profile each and generate recommendations for additional sales; the same input can update inventories by stock-keeping-unit and better manage shipping across the supply chain; sensor data can also track misplaced inventory in stores and warehouses.

With respect to velocity, sales data can be real-time, as well as aggregated to hourly, daily, weekly, and monthly form to support marketing decisions. Inventory data obtained in real-time can be aggregated to hourly or monthly updates. Location and time information can be organized to manage the supply chain.

Variety is magnified in this context by sales events from cash registers in brick-and-mortar locations, along with Internet sales, wholesale activity, international activity, and activity by competitors. All of this information can be combined with social media monitoring to better profile customers by profile. Inventory activity can be monitored by type of outlet as well as by vendor. Sensor-obtained data can be traced by workers involved, paths used, and locations.

1.1 Computer Support Systems

Computer systems have been applied to support business decision making for decades [5]. When personal computers came out, they were used to provide analytic tools for specific problems (**decision support systems**) [6]. Commercial software firms (such as Execucom and Comshare) extended this idea to dedicated systems to serve executives by providing them the key data they were expected to be interested in at their fingertips (**executive support systems**). Another commercial application was **on-line analytic processing**, developing database spreadsheet software capable of providing reports on any of a number of available dimensions.

In a parallel universe, statisticians and students of artificial intelligence revolutionized the field of statistics to develop data mining, which when combined with database capabilities evolving on the computer side led to **business intelligence**. The quantitative side of this development is **business analytics**, focusing on providing better answers to business decisions based on access to massive quantities of information ideally in real-time (**big data**) .

Davenport [7] reviewed three eras of analytics (see Table 1.1). The first era involved business intelligence, with focus on computer systems to support human decision making (for instance, use of models and focused data on dedicated computer systems in the form of decision support systems). The second era evolved in the early 21st Century introducing big data, through internet and social media generation of masses of data. Davenport sees a third era in a data-enriched

Table 1.1 The evolution of big data

	Era (Davenport)	Specific meaning
Decision support	1970–1985	Data analysis to support decision making
Executive support	1980–1990	Data analysis by senior executives
Online analytic processing	1990–2000	Analysis of multidimensional tables
Business intelligence	1989–2005	Tools to support data-driven decisions, emphasizing reporting
Analytics	2005–2010	Statistical and mathematical modeling for decision making
Big data	2010-now	Large, unstructured, fast-moving data

environment where on-line real-time analysis can be conducted by firms in every industry. This is accomplished through new tools, using Hadoop clusters and NoSQL databases to enable data discovery, applying embedded analytics supporting cross-disciplinary data teams.

One source of all of this data is the Internet of Things. Not only do people send messages now cars, phones, and machines communicate with each other [8]. This enables much closer monitoring of patient health, to include little wristbands to monitor the wearer's pulse, temperature, blood pressure forwarded on to the patient's physician. How people ever survived until 2010 is truly a wonder. But it does indicate the tons of data in which a miniscule bit of important data exists. Of course, signals are sent only when critical limits are reached, just as vending machines can send signals to replenish stock at their critical limits. Monitors in homes can reduce electricity use, thus saving the globe from excessive warming. Cars can send signals to dealers about engine problems, so that they might send a tow truck to the location provided by the car's GPS. Insurance already advertise their ability to attach devices to cars to identify good drivers, a euphemism for detection of bad driving so that they can cancel policies more likely to call for claims.

1.2   Examples of Knowledge Management

There have been impressive accomplishments using big data. Google detected the SARS epidemic much sooner than the US health system [9]. The Harvard Medical School found Tweets to be as accurate as official reports in tracking cholera after the 2010 Haitian earthquake, and two weeks faster. Movie firms have found Tweets to be good at predicting box-office revenues.

Wu et al. [10] provided a knowledge management framework for the product lifecycle, to include classification of knowledge types:

- Customer knowledge – CRM focus in data mining terms;
- Development knowledge—product design involving engineering expertise;

- Production knowledge—knowledge of production processes;
- Delivery and Service knowledge—knowledge of the processes needed to serve customers.

Knowledge of customers is a classical customer profiling matter. The other three bullets are classical business process reengineering matters, often involving tacit knowledge which organizations generate in the form of their employees' expertise. Management of these forms of knowledge require:

- A mechanism to identify and access knowledge;
- A method for collaboration to identify who, how, and where knowledge is;
- A method to integrate knowledge for effectively making specific decisions.

Data can be found in statistics of production measures, which accounting provides and which industrial engineers (and supply chain managers) analyze for decision making. Knowledge also exists in the experience, intuition, and insight found in employees (tacit information). This tacit knowledge includes organizational value systems. Thus expression of such knowledge is only available through collaboration within organizations. With respect to knowledge management, it means that the factual data found in accounting records needs to be supplemented by expertise, and a knowledge management system is closely tied to the idea of business process mapping. Business process mapping in turn is usually expressed in the form of a flowchart of what decisions need to be made, where knowledge can be found, and the approval authority in the organizations control system.

Kellmereit and Obodovski view this brave new world as a platform for new industries, around intelligent buildings, long-distance data transmission, and expansion of services in industries such as health care and utilities. Humans and machines are contended to work best in tandem, with machines gathering data, providing analytics, and applying algorithms to optimize or at least improve systems while humans provide creativity. (On the other hand, computer scientists such as Ray Kurzweil [11] expect machines to develop learning capabilities circe 2040 in the Great Singularity). Retail organizations (like Wal-Mart) analyze millions of data sets, some fed by RFID signals, to lower costs and thus serve customers better.

Use of all of this data requires increased data storage, the next link in knowledge management. It also is supported by a new data environment, allowing release from the old statistical reliance on sampling, because masses of data usually preclude the need for sampling. This also leads to a change in emphasis from hypothesis generation and testing to more reliance on pattern recognition supported by machine learning. A prime example of what this can accomplish is **customer relationship management**, where every detail of company interaction with each customer can be stored and recalled to analyze for likely interest in other company products, or management of their credit, all designed to optimize company revenue from every customer.

Knowledge is defined in dictionaries as the expertise obtained through experience or education leading to understanding of a subject. Knowledge acquisition refers to the processes of perception, learning, and reasoning to capture, structure, and represent knowledge from all sources for the purpose of storing, sharing, and implementing this knowledge. Our current age has seen a view of a knowledge being used to improve society.

Knowledge discovery involves the process of obtaining knowledge, which of course can be accomplished in many ways. Some learn by observing, others by theorizing, yet others by listening to authority. Almost all of us learn in different combinations of these methods, synthesizing different, often conflicting bits of data to develop our own view of the world. Knowledge management takes knowledge no matter how it is discovered and provides a system to provide support to organizational decision making.

In a more specific sense, knowledge discovery involves finding interesting patterns from data stored in large databases through use of computer analysis. In this context, the term **interesting** implies non-trivial, implicit, previously unknown, easily understood, useful and actionable knowledge. **Information** is defined as the patterns, correlations, rules, or relationships in data providing knowledge useful in decision making.

We live in an age swamped with data. As if satellite feeds of weather data, military intelligence, or transmission of satellite radio and television signals weren't enough, the devices of the modern generation including Twitter, Facebook, and their many competitors flood us with information. We sympathize with the idea that parents can more closely monitor their children, although whether these children will develop normally without this close supervision remains to be seen. But one can't help wonder how many signals containing useless information clutter up the lives of those with all of these devices.

1.3 Data Mining Forecasting Applications

Knowledge management consists of the overall field of human knowledge (epistemology) as well as means to record and recall it (computer systems) and quantitative analysis to understand it (in business contexts, business analytics). There are many applications of quantitative analysis, falling within the overall framework of the term business analytics. Analytics has been around since statistics became widespread. With the emergence of computers, we see three types of analytic tools. **Descriptive** analytics focus on reports of what has happened. Statistics are a big part of that. **Predictive** analytics extend statistical and/or artificial intelligence to provide forecasting capability. It also includes classification modeling, applying models to suggest better ways of doing things, to include identification of the most likely customer profiles to send marketing materials, or to flag suspicious insurance claims, or many other applications. **Diagnostic** analytics can apply analysis to sensor input to direct control systems automatically. This is especially useful in mechanical or chemical environments where speed and safety considerations make it attractive to replace human monitors with automated systems as much as possible. It can lead to some problems, such as bringing stock markets to their knees for short periods (until humans can regain control). **Prescriptive** analytics applies quantitative models to optimize systems, or at least to identify improved systems. Data mining includes descriptive and predictive modeling. Operations research includes all four. This book focuses on the forecasting component of predictive modeling, with the classification portion of prescriptive analytics demonstrated.

There are many applications of predictive analytics involving forecasting models in practically every field of human activity. Governments use automated systems and manage raw data from multiple streams to include monitoring mortgage insurance fund performance [12]. This application predicted the number of defaulted loans in a federal credit agency loan portfolio, including a risk-ranking model to aid regulatory compliance enforcement. In the chemical process industry, exponential smoothing along with other statistical analysis has been applied to forecast demand more effectively [13]. In the field of mechanical engineering, industrial processing has been used to deal with skilled worker shortage to monitor heat exchanger performance, detecting when maintenance activity is appropriate [14]. The flood of real-time data generated was programmed to display to humans in a timely and easy-to-recognize fashion, part of the predictive analytics software. In the field of human resources management, predictive workforce analytics has been developed to apply multivariate regression models to understand and manage employee turnover [15]. Predictive analytics has been applied in construction operations to predict workplace industries, monitoring staff hours, training, production schedules, video monitoring and other sensor information to identify risk at specific locations, along with suggestions for mitigation [16]. Predictive analytics has also been applied pedagogically to monitor student grade and success likelihood in classroom environments [17]. Of the many applications of predictive analytics to the health industry, models have applied regression and functional approximation to more accurately predict clinical patient length of stay [18].

1.4   Summary

The primary purpose of knowledge management is to wade through all of this noise to pick out useful patterns. That is data mining in a nutshell. Thus we view knowledge management as:

- Gathering appropriate data

  - Filtering out noise

- Storing data (DATABASE MANAGEMENT)
- Interpret data and model (DATA MINING)

  - Generate reports for repetitive operations
  - Provide data as inputs for special studies.

Descriptive modeling are usually applied to initial data analysis, where the intent is to gain initial understanding of the data, or to special kinds of data involving relationships or links between objects. The crux of data mining modeling is classification, accomplished by logistic regression, neural network, and decision tree prescriptive models. This book presents simple explanations and demonstration of some forecasting tools. Chapter 3 covers simple linear regression. This is extended in Chap. 4 to causal regression. Chapter 5 covers regression trees. Chapter 6 describes and demonstrates autoregressive integrated moving-average (ARIMA) and generalized autoregressive conditional heteroscedasticity (GARCH) models. Chapter 7 provides a quick overview of classification modeling. Chapter 8 reviews how these different kinds of predictive models provide value in the era of big data.

References

1. Davenport TH (2014) Big data at work. Harvard Business Review Press, Boston
2. Rothberg HN, Erickson GS (2005) From knowledge to intelligence: creating competitive advantage in the next economy. Elsevier Butterworth-Heinemann, Woburn, MA
3. Manyika J, Chui M, Brown B, Bughin J, Dobbs R, Roxburgh C, Byers HA (2011) Big data: the next frontier for innovation, competition and productivity. McKinsey Global Institute, New York
4. Waller MA, Fawcett SE (2013) Data science, predictive analytics, and big data: a revolution that will transform supply chain design and management. Journal of Business Logistics 34 (2):77–84
5. Olson DL, Courtney JF Jr (1992) Decision support models and expert systems. MacMillan Publishing Co., New York
6. Sprague RH, Carlson ED (1982) Building effective decision support systems. Prentice-Hall, Englewood Cliffs, NJ
7. Davenport TH (2013) Analytics 3.0. Harvard Bus Rev 91(12):64–72
8. Kellmereit D, Obodovski D (2013) The silent intelligence: the internet of things. DnD Ventures, San Francisco
9. Brynjolfsson E, McAfee A (2014) The second machine age: work, progress, and prosperity in a time of brilliant technologies. W.W. Norton & Co., New York
10. Wu ZY, Ming XG, Wang YL, Wang L (2014) Technology solutions for product lifecycle knowledge management: framework and a case study. Int J Prod Res 52(21):6314–6334
11. Kurzweil R (2000) The age of spiritual machines: when computers exceed human intelligence. Penguin Books, New York
12. Lee AJ (2015) Predictive analytics: the new tool to combat fraud, waste and abuse. J Gov Financ Manag 64(2):12–16
13. Blackburn R, Lurz K, Priese B, Göb R, Darkow I-L (2015) A predictive analytics approach for demand forecasting in the process industry. Int Trans Oper Res 22(3):407–428
14. Olsen T (2016) Improving heat exchanger operation with predictive analytics. Process Heating 23(6):20–23
15. Hirsch W, Sachs D, Toryfter M (2015) Getting started with predictive workforce analytics. Workforce Solutions Rev 6(6):7–9
16. Parsons J (2016) Predictive analytics: the key to injury prevention? Eng New-Record CBQ10-CBA-12. 22 Feb 2016
17. Rahal A, Zainuba M (2016) Improving students' performance in quantitative courses: the case of academic motivation and predictive analytics. Int J Manag Educ 14(1):8–17
18. Harris SL, May JH, Vargas LG (2016) Predictive analytics model for healthcare planning and scheduling. Eur J Oper Res 253(1):121–131

Chapter 2
Data Sets

Data comes in many forms. The current age of big data floods us with numbers accessible from the Web. We have trading data available in real time (which caused some problems with automatic trading algorithms, so some trading sites impose a delay of 20 min or so to make this data less real-time). Wal-Mart has real-time data from its many cash registers enabling it to automate intelligent decisions to manage its many inventories. Currently a wrist device called a Fitbit is very popular, enabling personal monitoring of individual health numbers, which have the ability to be shared in real-time with physicians or ostensibly EMT providers. The point is that there is an explosion of data in our world.

This data has a lot of different characteristics. Wal-Mart might have some data that is very steady, with little trend. Things like demand for milk, which might have a very predictable relationship to population. You might need to know a bit more about age group densities, but this kind of data might be big if it is obtained from grocer cash registers across a metropolitan region, but is likely to be fairly stable. Other kinds of data follow predictable theories. A field is a science when it is understood to the extent that it can be mathematically modelled. (That's why some deny that economics is a science—it can be mathematically modelled, but it is often not correct in its predictions.) Other types of data behave in patterns impossible to define over a long term.

Real data, especially economic data, is very difficult to predict [1]. Examples include the Internet bubble, the Asian financial crisis, global real estate crises, and many others just in the past 15 years. Nobody predicted failure of Long-Term Capital Management in the 1990s, nor Enron and WorldCom a few years later, nor Lehman Brothers, Northern Rock, and AIG around the global real estate crisis. Physics and engineering benefit from consistent behavior of variables, and can be

Electronic supplementary material The online version of this chapter (doi:10.1007/978-981-10-2543-3_2) contains supplementary material, which is available to authorized users.

modeled to the extent that rockets can be sent to the moon and Mars and be expected to get there. Casinos have games that, given the assumption that they are fair, have precise probabilities. But when human choice gets involved, systems usually get too complex to accurately predict.

The field of complexity [2] was discussed by John Holland, and has been considered by many others. Holland viewed complicate mechanisms as those that could be designed, predicted, and controlled, such as automobile engines or pacemakers, or refineries. Complicated mechanisms can consist of a very large number of interacting parts, but they behave as their science has identified. Complex systems, on the other hand, cannot be completely defined. Things such as human immunology, global weather patterns, and networks such as Facebook and LinkedIn consist of uncountable components, with something like order emerging, but constantly interacting making precise prediction elusive. Ramo [3] described how this emerging world has grown in complexity to the point that a new paradigm shift has occurred that will change the world. People have Internet connections that can no longer be controlled.

Big data thus consists of an infinite stream of constantly generated data. Prices of stocks change as buyers and sellers interact. Thus trading data is instantaneous. Data mining tools described in this book can be applied to this real-time data, but in order to describe it, we need to look at it from a longer perspective. We will focus on data related to business, specifically Chinese trading data. While we will be looking at a small set of monthly data, that is to make it more understandable. We will also look at daily data, but the methods presented can be applied to data on any scale. Applying it to real-time data makes it big data.

We will consider three data series. The first is the price of gold, which represents a view of investor conservatism. The second is the price of Brent crude oil. We will use monthly data for the period 2001 through April 2016 for this overview. The third data set consists of four stock indices: the S&P 500 represents conservative US investor views; NYSE is the New York Stock Exchange; Eurostoxx is a European stock index; MXCN is the Morgan Stanley Capital Index for a composite of Chinese investment options will be presented in Chap. 4.

2.1 Gold

The price of gold in US dollars per troy ounce is shown in Fig. 2.1.

We view the price of gold as an indicator of investor conservatism—a high price of gold indicates concern over investment opportunities. The time series shows a very steady behavior from the beginning of 2001 until a bit of a jump in 2006, then on to a peak in early 2008, a slight drop with the realty bubble in 2008 followed by an unsteady rise to nearly $1800 per ounce in 2011, followed by a drop from late 2012 to its hovering around $1200 per ounce in early 2016. Thus the data from 2001 through 2005 demonstrates linearity, the kind of data that linear regression can forecast well. That is, until it starts to display nonlinear behavior, as it does

Fig. 2.1 Gold

starting in 2006. Figure 2.2 shows this linear trend, explaining 0.957 of the change in the price of gold by the formula Trend = 254.9194 + 3.787971 *Time (with Time = 1 in January 2001).

For the entire range 2001 through 2015 the linear regression for this data explains nearly 80 % of the change. This time series has interesting cyclical behavior that might lead to improvement over the linear model. It is viewed as an input to forecasting other variables, such as oil or stock indices.

In general, economic time series tend to display a great deal of nonlinearity, consisting of a series of switchbacks and discontinuities. The linearity displayed by the price of gold from 2001 through 2005 was thus atypical. The factor that led to nonlinearities in 2006 was probably the precursor to the high levels of risk perceived in the economy, especially in real estate, that led to near-economic collapse in 2008.

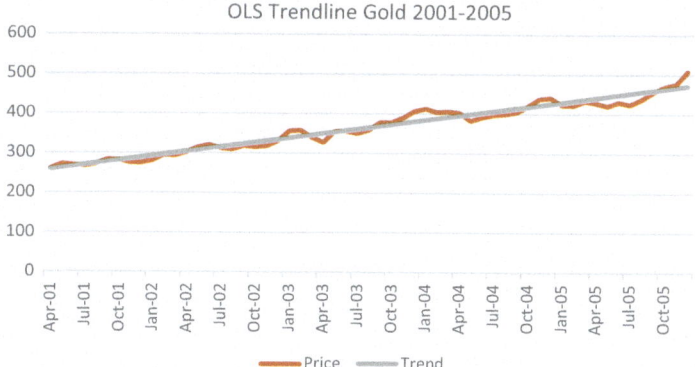

Fig. 2.2 OLS trendline gold 2001–2005

2.2 Brent Crude

The price of crude oil reflects reliance upon fossil fuel for transportation. There are many forces that work on this time series, including general economic activity, control by producers (OPEC), alternative sources (non-fossil fuels, new production from fracking), and the policies of many governments. This series was very linear through 1972, but has had very interesting behavior since 1973, when OPEC became active in controlling member production. There have been many governmental (and OPEC) policy actions (to include wars) that have seen high degrees of nonlinearity in the price of crude oil. Another major technological change was the emergence of fracking around 2006. Figure 2.3 displays the time series for Brent North Sea oil, one source of crude oil.

In 2001 the thinking was that peak oil had been reached, so that the price could be expected to rise as supply was exhausted. You can see a slight increase through the middle of 2006. This was followed by a noted drop in price, followed by a steep increase until the middle of 2008 as the global economy surged. The sudden drop in 2008 came from the bursting of the real estate bubble, seeing a dramatic decrease from a high of about $135 per barrel in early 2008 to just over $40 per barrel later in the year. The price came back, reaching over $120 in 2011, and remaining in the 100–120 range until the middle of 2014, when the price sharply dropped to around $50 per barrel, and after a short recovery dropped again into the $30s.

As a time series, crude oil has very volatile behavior. Time itself thus does not explain why it changes. We will run a linear regression, which explains just over 40 percent of the change by r-squared measure. But clearly there are other factors explaining the price of oil, such as general economic activity as well as perceptions of risk. We will demonstrate the behavior of ARIMA and GARCH models that take advantage of cyclical time series content along with trend. We will demonstrate multiple regressions for this time series as well. Keep in mind that our purpose is to demonstrate forecasting methods, not to explain why this particular series behaves as it has.

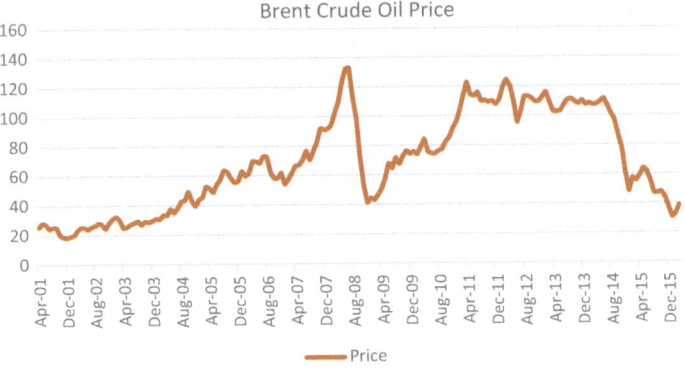

Fig. 2.3 Brent crude oil

2.3 Stock Indices

The S&P 500 is a relatively conservative stock index based upon a market basket of 500 stocks listed on the New York stock exchange. Its monthly behavior is displayed in Fig. 2.4.

This index displayed negative returns from the beginning of 2001 through 2003, recovered slightly through 2007, suffered a severe decline through the middle of 2009, after which it has slowly recovered to a new peak in early 2015 and has been fairly steady since. We view this series as a measure of relatively conservative investor confidence. The simple linear regression indicates that time by itself explains 52 % of its change, but there are obvious cyclical behaviors that might provide greater forecasting accuracy from ARIMA or GARCH models.

We have other indices as well, to include the New York Stock Exchange index, a broader measure of which the S&P is a component. Results are very similar to the S&P 500, as shown in Fig. 2.5.

Fig. 2.4 S&P

Fig. 2.5 NYSE

We include Eurostoxx (see Fig. 2.6) reflecting European stock investment and the Shenzhen index representing Chinese stock investment.

Eurostoxx can be seen to have not recovered from either their 2001 data set start, or the peak before the 2008 real estate bubble collapse. Yet it has seen something of an increase since the end of 2012. The linear regression model provides an r-square measure of only 0.045, or 4.5 %. Clearly other factors explain its performance beyond time.

We will model the monthly price of the MSCI (Morgan Stanley Capital International) China Index. A description of this time series is found at www.MSCI. com/China. This index was launched in October 1995, with back-tested data calculated. MSCI is intended to reflect Chinese mainland equity markets, to include trade on the Shanghai, Shenzhen, and Hong Kong exchanges, of both state-owned and non-state-owned shares. Monthly data for the period January 2001 through April 2016 yielded the time series shown in Fig. 2.7.

Fig. 2.6 Eurostoxx

Fig. 2.7 MSCI China

2.4   Summary

The data we have presented are economic time series, which usually involve high levels of nonlinearity over a long enough period of time. Sometimes these series display stability, such as the price of gold in the period 2000 through 2005. In that case, simple models, as presented in the next chapter, work quite well. These include moving average and ordinary least squares regression. But economic time series usually become more interesting, and thus harder to forecast. One approach is to build causal regression (covered in Chap. 4) or regression tree models (Chap. 5). Autoregressive moving average (ARIMA) and generalized autoregressive conditional heteroscedasticity (GARCH) models are more powerful tools to reflect cyclical behavior, and will be covered in subsequent chapters.

References

1. Makridakis S, Taleb N (2009) Decision making and planning under low levels of predictability. Int J Forecast 25:716–733
2. Holland JH (1995) Hidden order: how adaptation builds complexity. Perseus Books, Cambridge
3. Ramo JC (2016) The seventh sense: power, fortune and survival in the age of networks. Little, Brown and Company, NY

Chapter 3
Basic Forecasting Tools

We will present two fundamental time series forecasting tools. Moving average is a very simple approach, presented because it is a component of ARIMA models to be covered in a future chapter. Regression is a basic statistical tool. In data mining, it is one of the basic tools for analysis, used in classification applications through logistic regression and discriminant analysis, as well as prediction of continuous data through ordinary least squares (OLS) and other forms. As such, regression is often taught in one (or more) three-hour courses. We cannot hope to cover all of the basics of regression. However, we here present ways in which regression is used within the context of data mining.

3.1 Moving Average Models

The idea behind moving average models is quite simply to take the n last observations and divide by n to obtain a forecast. This can and has been modified to weight various components, but our purpose is to simply demonstrate how it works. We use the monthly Brent crude oil data to demonstrate. Table 3.1 shows the first year of monthly observations, which is then used to generate two-period, three-period, and twelve-period forecasts. Each forecast thus has the same number of opportunities to be wrong.

The observations for January 2002 are obtained by taking the n prior Prices and dividing by n. For the two-period model, this would be $(18.94 + 18.60)/2 = 18.77$. For the three-period model, $(20.48 + 18.94 + 18.60)/3 = 19.34$. For the twelve-period model all twelve prices from January 2001 through December 2001 are summed and divided by 12, yielding 24.42. The actual price in January 2002 was 19.48, and the squared error of each forecast is the difference between forecast and actual squared. We could use other metrics such as absolute value or even greatest error in a column, but choose squared error to compare with regression output below.

© Springer Science+Business Media Singapore 2017
D.L. Olson and D. Wu, *Predictive Data Mining Models*,
Computational Risk Management, DOI 10.1007/978-981-10-2543-3_3

Table 3.1 Moving average forecasts—brent crude

Month	Time	Brent	2pd	SqdError	3pd	SqdError	12pd	SqdError
	1	25.62						
	2	27.5						
	3	24.5						
Apr-01	4	25.55						
May-01	5	28.45						
Jun-01	6	27.72						
Jul-01	7	24.54						
Aug-01	8	25.67						
Sep-01	9	25.54						
Oct-01	10	20.48						
Nov-01	11	18.94						
Dec-01	12	18.6						
Jan-02	13	19.48	18.77	0.50	19.34	0.01	24.42	24.46
Feb-02	14	20.29	19.04	1.56	19.01	1.65	23.91	13.13
Mar-02	15	23.69	19.89	14.48	19.46	17.92	23.31	0.14

The average squared errors over the entire data set (January 2002 through April 2016) were:

2-Period moving average 61.56
3-period moving average 89.00
12-period moving average 275.84.

These outcomes reflect that with this relatively volatile time series, the closer to the present you are, the more accurate the forecast. Moving average has the advantage of being quick and easy. As a forecast, it only extends one period into the future, which is a limitation. Using the data through April 2016, the forecasts for May 2016 were:

2-Period moving average 42.45
3-period moving average 39.37
12-period moving average 46.35.

Keep in mind that if there was a seasonal cycle, the 12-period moving average might be a very good forecast. But Brent crude oil doesn't seem to have that much of a seasonal component. Nonetheless, moving average can be a valuable component of ARIMA models, which we will cover later.

3.2 Regression Models

Regression is used on a variety of data types. If data is time series, output from regression models is often used for forecasting. Regression can be used to build predictive models for other types of data. Regression can be applied in a number of different forms. The class of regression models is a major class of tools available to support the Modeling phase of the data mining process.

Probably the most widely used data mining algorithms are data fitting, in the sense of regression. Regression is a fundamental tool for statistical analysis to characterize relationships between a dependent variable and one or more independent variables. **Regression** models can be used for many purposes, to include explanation and prediction. Linear and **logistic regression** models are both primary tools in most general purpose data mining software. Nonlinear data can sometimes be transformed into useful linear data and analyzed with linear regression. Some special forms of nonlinear regression also exist.

Ordinary least squares regression (OLS) is a model of the form:

$$Y = \beta_0 + \beta_1 X_1 + \beta_2 X_2 + \cdots + \beta_n X_n + \varepsilon$$

where Y is the dependent variable (the one being forecast)

X_n are the n independent (explanatory) variables
β_0 is the intercept term
β_n are the n coefficients for the independent variables
ε is the error term.

OLS regression is the straight line (with intercept and slope coefficients β_n) which minimizes the sum of squared error terms ε_i over all i observations. The idea is that you look at past data to determine the β coefficients which worked best. The model gives you the most likely future value of the dependent variable given knowledge of the X_n for future observations. This approach assumes a linear relationship, and error terms that are normally distributed around zero without patterns. While these assumptions are often unrealistic, regression is highly attractive because of the existence of widely available computer packages as well as highly developed statistical theory. Statistical packages provide the probability that estimated parameters differ from zero.

We can apply regression to the problem of extending a trend line. We will use monthly data for the price of Brent crude oil to demonstrate. The dependent variable (Y) is the monthly average price of Brent crude over the period 2001 through March 2016. The independent variable (X) is time, an index of weeks beginning with 1 and ending at 180, the last available observation. Figure 3.1 displays this data.

This data is quite erratic, and notably nonlinear. Ordinary least squares (OLS) regression fits this data with the straight line that minimizes the sum of squared

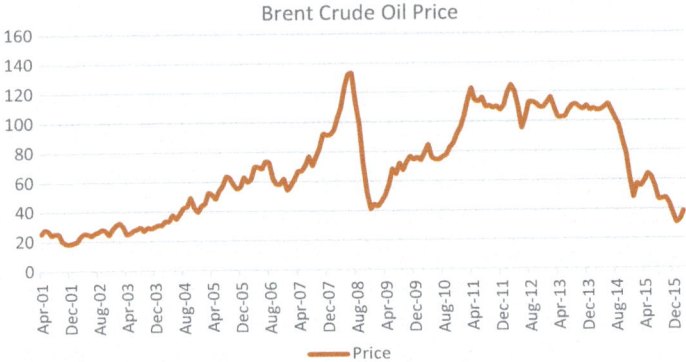

Fig. 3.1 Graph of Brent crude

error terms. Given the data's nonlinearity, we don't expect a very good fit, but the OLS model does show average trend. Here the model is:

$$Y = \beta_0 + \beta_1 X + \varepsilon$$

where Y is Requests and X is week.

The regression output from Excel for our data is shown in Table 3.2.

Table 3.2 Regression output for time series data

Summary output						
Regression statistics						
Multiple R	0.658909					
R square	**0.434161**					
Adjusted R square	0.430982					
Standard error	24.32224					
Observations	180					
ANOVA						
	df	SS	MS	F	Significance F	
Regression	1	80795.2	80795.2	136.5773	8.79E-24	
Residual	178	105299.7	591.5713			
Total	179	186094.9				
	Coefficients	Standard error	t Stat	*P*-value	Lower 95 %	Upper 95 %
Intercept	**31.99853**	3.640905	8.788619	1.26E-15	24.81364	39.18342
Time	**0.407738**	0.034889	11.68663	8.79E-24	0.338888	0.476588

This output provides a great deal of information. We will discuss regression statistics, which measure the fit of the model to the data, below. ANOVA information is an overall test of the model itself. The value for *Significance F* gives the probability that the model has no information about the dependent variable. Here, 8.79E-24 is practically zero (move the decimal place 24 digits to the left, resulting in a lot of zeros). MS for the Residual is the mean squared error, which can be compared to the moving average forecasts above. Here the value of 591.57 is worse than any of the three moving average forecasts we calculated. Finally, at the bottom of the report, is what we were after, the regression model.

$$Price = 31.99853 + 0.407738 \times Time$$

where Time = 1 January 2001, and Time = 180 March 2016.

This enables us to predict the price of Brent crude into the future. It is tempting to extrapolate this model into the future, which violates the assumptions of regression model. But extrapolation is the model's purpose in prediction. Still, the analyst needs to realize that the model error is expected to grow the farther the model is projected beyond the data set upon which it was built. To forecast, multiply the time index by 0.407738 and add 31.99853. The forecasts for months 181 through 185 are given in Table 3.3.

The graphical picture of this model is given in Fig. 3.2.

Table 3.3 Time series forecasts from regression model

Time	Prediction
181	105.80
182	106.21
183	106.61
184	107.02
185	107.43

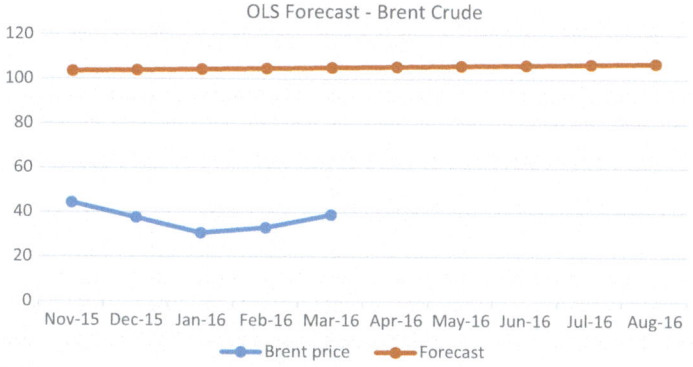

Fig. 3.2 Graph of time series model and forecasts

Clearly this OLS forecast is way off. It predicts a price a bit more than $100 per barrel, when the actual is close to $40 per barrel. But this price was over $100 as recently as August 2014, and Fig. 3.1 shows how volatile this time series is.

3.3 Time Series Error Metrics

The classical tests of regression models are based on the assumption that errors are normally distributed around the mean, with no patterns. The basis of regression accuracy are the residuals, or difference between prediction and observed values. Residuals are then extended to a general measure of regression fit, R-squared.

SSE: The accuracy of any predictive or forecasting model can be assessed by calculating the sum of squared errors (SSE). In the regression we just completed, SSE is 105299.7. Each observation's residual (error) is the difference between actual and predicted. The sign doesn't matter, because the next step is to square each of these errors. The more accurate the model is, the lower its SSE. An SSE doesn't mean much by itself. But it is a very good way of comparing alternative models, if there are equal opportunities for each model to have error.

R^2: SSE can be used to generate more information for a particular model. The statistic R^2 is the ratio of variance explained by the model over total variance in the data. Total squared values (186094.9 in our example) is explained squared dependent variable values (80795.2 in our example) plus SSE (105299.7 in our example). To obtain R^2, square the predicted or forecast values of the dependent variable values, add them up (yielding MSR), and divide MSR by (MSR + SSE). This gives the ratio of change in the dependent variable explained by the model (80795.2/ 186094.9 = 0.434161 in our example). R^2 can range from a minimum of 0 (the model tells you absolutely nothing about the dependent variable) to 1.0 (the model is perfect).

$$R^2 = \frac{SST - SSE}{SST}$$

where

SST   is the sum of squared deviations of the dependent variable from its own mean,

SSE   is the sum of squared error (difference between actual dependent variable values and predicted or forecast values).

There are basic error metrics for general time series forecasts. The regression models from Excel report SSE as we have just seen. Another way to describe SSE is to take error for each observation, square these, and sum over the entire data set. Calculating the mean of SSE provides mean squared error (MSE). An alternative error metric often used is mean absolute deviation (MAD), which is the same thing except that instead of squaring errors, absolute values are taken. MAD is considered more robust than MSE, in that a very large error will affect MSE much more than MAD. But both provide useful means of comparing the relative accuracy of time series, given that they compare models over exactly the same data. A more general error metric is mean absolute percentage error. Here the error for each observation is calculated as a percentage of the actual observation, and averaging over the entire time series.

3.4   Seasonality

Seasonality is a basic concept—cyclical data often has a tendency to be higher or lower than average in a given time period. For monthly data, we can identify each month's average relative to the overall year. Using Brent crude oil for demonstration, we can calculate the average by month from our data set for January through December over the period 2001 through the end of 2015. Table 3.4 shows these averages.

The calculation is trivial—using the Average function in Excel for data by month, and the overall average for entire series (here labelled "year"). Note that there is a bit of bias because of trend, which we will consider in time series regression. Nonetheless, this data shows a bit of seasonality, as shown in Fig. 3.3.

It appears that July and August have higher prices on average. This could be because of some seasonal variation like more people consume gasoline to drive vehicles in the Summer. However, it also could be spurious. Given the relatively smooth nature of the curve in Fig. 3.1, however, there actually might be a seasonal component to the price of crude oil.

We can include seasonality into the regression against Time by using dummy variables for each month (0 if that observation is not the month in question—1 if it is). This approach requires skipping one time period's dummy variable or else the model would be over-specified, and OLS wouldn't work. We'll skip December. An extract of this data is shown in Table 3.5.

Table 3.4 Seasonality Indices by month—brent crude oil	Month	Average	SeasonIndex
	Jan	63.94	0.93
	Feb	66.25	0.96
	Mar	68.73	1.00
	Apr	70.63	1.03
	May	70.96	1.03
	June	71.48	1.04
	Jul	72.63	1.06
	Aug	72.00	1.05
	Sep	70.15	1.02
	Oct	68.17	0.99
	Nov	65.93	0.96
	Dec	64.22	0.93
	year	68.76	

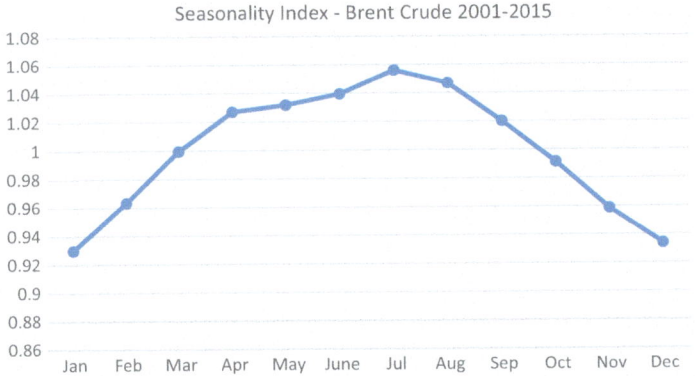

Fig. 3.3 Seasonality indices by month

Table 3.5 Seasonal regression data extract—brent crude

Month	Time	Brent	Time	Jan	Feb	Mar	Apr	May	Jun	Jul	Aug	Sep	Oct	Nov
Jan01	1	25.62	1	1	0	0	0	0	0	0	0	0	0	0
Feb01	2	27.5	2	0	1	0	0	0	0	0	0	0	0	0
Mar01	3	24.5	3	0	0	1	0	0	0	0	0	0	0	0
Apr01	4	25.55	4	0	0	0	1	0	0	0	0	0	0	0
May01	5	28.45	5	0	0	0	0	1	0	0	0	0	0	0
Jun01	6	27.72	6	0	0	0	0	0	1	0	0	0	0	0

The Excel output for this model is given in Table 3.6.

Note that R Square increased from 0.434 in Table 3.2 to 0.536, which sounds great. Adjusted R Square rose from 0.431 to 0.502, more convincing. Note however that while Time is still significant, none of the dummy variables are. The model is now:

$$\text{Brent} = 20.76 + 0.45 \, \text{Time} + \text{dummy variable by month}$$

The nice thing about this regression is that there is no error in guessing future variable values. However, the forecasts for 2016 are shown in Table 3.7.

The seasonal model has more content, and statistically looks stronger. But it is even worse than the simple time regression here. Both are weak and pathetic forecasts. That is clearly because Fig. 3.1 shows that this data is very nonlinear, and the Time Regression assumes a straight line. Given no additional knowledge, that's about as good a bet as any. Here seasonality is weak, but provides more information. In this case, the forecast gets even worse. In reality, the price of oil undoubtedly depends a great deal more on complex economic factors, and even more on the decisions of those who Saudi and Russian oil ministers work for.

Table 3.6 Seasonal regression model—brent crude

Summary output						
Regression statistics						
Multiple R	0.731838					
R square	0.535587					
Adjusted R square	0.502216					
Standard error	22.86715					
Observations	180					
ANOVA						
	df	SS	MS	F	Significance F	
Regression	12	100708.4	8392.366	16.04946	2.88E-22	
Residual	167	87325.39	522.9066			
Total	179	188033.8				
	Coefficients	Standard error	t Stat	P-value	Lower 95 %	Upper 95 %
Intercept	20.7569	6.694816	3.100444	0.002269	7.539524	33.97429
Time	0.452761	0.032875	13.77236	2.55E-29	0.387858	0.517665
Jan	4.696375	8.35773	0.56192	0.574924	−11.804	21.1968
Feb	6.557614	8.356372	0.784744	0.433715	−9.94013	23.05536
Mar	8.586186	8.355143	1.027653	0.305599	−7.90913	25.0815
Apr	10.02609	8.354043	1.200148	0.231781	−6.46705	26.51924
May	9.902663	8.353073	1.185511	0.237499	−6.58857	26.39389
Jun	9.977235	8.352232	1.194559	0.233953	−6.51233	26.4668
Jul	10.67447	8.35152	1.278147	0.202971	−5.81369	27.16264
Aug	9.588379	8.350938	1.14818	0.252536	−6.89864	26.07539
Sep	7.284951	8.350485	0.872399	0.384244	−9.20117	23.77107
Oct	4.856189	8.350161	0.581568	0.561642	−11.6293	21.34167
Nov	2.157428	8.349967	0.258376	0.796435	−14.3277	18.64253

Table 3.7 Comparative forecasts

Time	Month	Time regression	Seasonal regression	Actual
181	Jan 2016	105.80	107.40	30.80
182	Feb 2016	106.21	109.71	33.20
183	Mar 2016	106.61	112.20	39.07
184	Apr 2016	107.02	114.09	45.83
185	May 2016	107.43	114.42	?
186	Jun 2016	107.84	114.95	?

(And they would point to the Seven Sisters and other oil companies, not to mention the U.S. and Chinese administrations.)

We can demonstrate error metrics with these forecasts. Table 3.8 shows calculations.

Table 3.8 Error calculations

Time	Actual	Time Reg	Abs error	Squared error	% error	Seas Reg	Abs error	Squared error	% error
181	30.80	105.80	75.00	5625.00	243.51	107.40	76.60	5867.56	248.70
182	33.20	106.21	73.01	5330.46	219.91	109.71	76.51	5853.78	230.45
183	39.07	106.61	67.54	4561.65	172.87	112.20	73.13	5348.00	187.18
184	45.83	107.02	61.19	3744.22	133.52	114.09	68.26	4659.43	148.94
Mean			**69.19**	**4815.33**	**192.45**		73.63	5432.19	203.82

Absolute error is simply the absolute value of the difference between actual and forecast. Squared error squares this value (skipping applying the absolute function, which would be redundant). Absolute percentage error divides absolute error by actual. The metric is the mean for each of these measures. In this case, all three metrics indicate that the Time regression is more accurate than the Seasonal regression. It often happens that these three error metrics yield the same results, but not always. Squared error magnifies extreme cases. Mean absolute percentage error allows comparison across forecasts with different scales.

3.5 Daily Data

This time series is available in daily form as well. The source from the Web was:

Europe Brent Spot Price FOB
http://www.eia.gov/dnav/pet/hist/LeafHandler.ashx?n=PET&s=rbrte&f=D
07:59:35 GMT-0500 (Central Standard Time)
Data Source: Thomson Reuters

Figure 3.4 shows this display for the period January 2000 through mid-May 2016.

Table 3.9 gives moving average forecasts for Brent crude oil close using daily data.

Again, it can be seen that shorter-range moving average forecasts have much less error than longer-range forecasts, which is generally true for highly volatile time series.

The regression model from Excel for this data is shown in Table 3.10.

The model is thus: Brent close = 26.39912 + 0.018669 × time (with 16 May 2016 time = 4157). Mean squared error was 567.10, better than the error from the monthly data regression, but still quite a bit worse than the moving average forecasts. This daily trend line explained almost 47 % of the price. The trend line is displayed in Fig. 3.5.

Fig. 3.4 Daily price close—brent crude

Table 3.9 Moving average forecasts—brent crude

Date	BrentPrice	5-day MA	Error	20-day MA	Error	250-day MA	Error
5/3/2016	43.09	45.03	1.94	40.49	2.60	45.43	2.34
5/4/2016	43.08	44.86	1.78	41.20	1.88	45.34	2.26
5/5/2016	44.39	44.65	0.26	41.74	2.65	45.25	0.86
5/6/2016	44.6	44.40	0.20	42.37	2.23	45.16	0.56
5/9/2016	42.43	44.20	1.77	42.70	0.27	45.08	2.65
5/10/2016	44.01	43.52	0.49	42.83	1.18	44.99	0.98
5/11/2016	46.08	43.70	2.38	42.93	3.15	44.91	1.17
5/12/2016	46.43	44.30	2.13	43.06	3.37	44.85	1.58
5/13/2016	47.05	44.71	2.34	43.32	3.73	44.77	2.28
5/16/2016	48.49	45.20	3.29	43.75	4.74	44.70	3.79
Avg Err		46.41	1.58	44.17	4.20	44.65	10.76

3.6 Change in Daily Price

We also can explore what trends might exist in daily closing price. We transform the data by:

$$\{\text{Current price}/\text{Previous day's price}\} - 1$$

Running a regression against time, we obtain the Excel regression output in Table 3.11.

This model has practically zero trend, indicated by the insignificant beta coefficient for Time, the practically zero r-squared value, and the plot shown in Fig. 3.6.

Table 3.10 Excel output for daily brent close

Summary output						
Regression statistics						
Multiple R	0.685198					
R square	**0.469497**					
Adj R square	0.469369					
Standard error	23.81377					
Observations	4156					
ANOVA						
	df	SS	MS	F	Significance F	
Regression	1	2084813	2084813	3676.3	0	
Residual	4154	2355715	567.0956			
Total	4155	4440528				
	Coefficients	Standard error	t Stat	*P*-value	Lower 95 %	Upper 95 %
Intercept	**26.39912**	0.738922	35.72652	5.2E-244	24.95044	27.8478
Time	**0.018669**	0.000308	60.6325	0	0.018065	0.019272

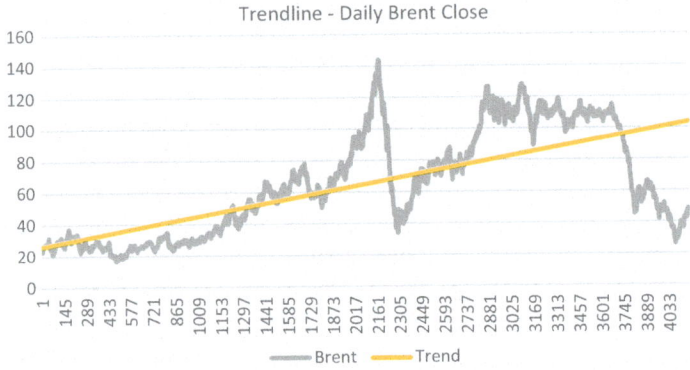

Fig. 3.5 Simple OLS regression daily brent crude

Despite this terrible fit, the mean squared error is much lower than we found in the forecasts for Brent crude price. This is because the scale of numbers in the data was much smaller. You can only compare MSE when the data has the same scale. The purpose was to show how simple OLS regression can be obtained. Looking at Brent Crude, there clearly is an upward trend in this data, but that trend explains less than half of the change. As you might expect, other factors are important,

Table 3.11 Excel regression of daily price change

Summary output						
Regression statistics						
Multiple R	0.020245					
R square	**0.00041**					
Adj R square	0.000169					
Standard error	0.022847					
Observations	4156					
ANOVA						
	df	SS	MS	F	Significance F	
Regression	1	0.000889	0.000889	1.70325	0.191936	
Residual	4154	2.168398	0.000522			
Total	4155	2.169287				
	Coefficients	Standard error	t Stat	*P*-value	Lower 95 %	Upper 95 %
Intercept	**0.001232**	0.000709	1.737964	0.082291	−0.00016	0.002622
Time	**−3.9E-07**	2.95E-07	−1.30509	0.191936	−9.6E-07	1.94E-07

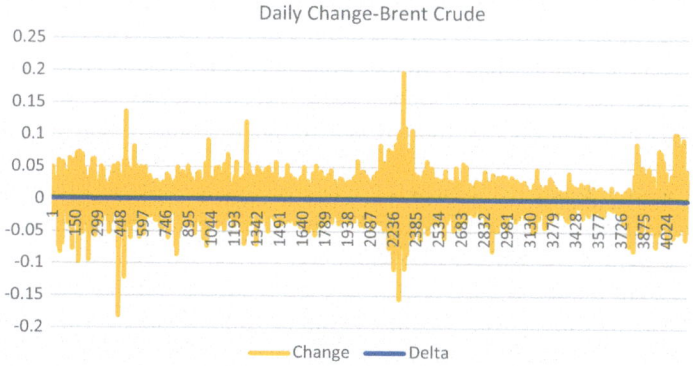

Fig. 3.6 Plot of daily change in brent crude price

which we will address in the next chapter. The daily change in price is practically zero, although the very slightly negative beta coefficient (the *p*-value of 0.19 indicates insignificance at any reasonable level) shows that it usually drops. Together with the positive overall trend, that indicates that when the price rises, it jumps more than it drops on those days that it declines.

3.7   Software Demonstrations

For ordinary least squares regression, Excel was demonstrated above. The only limitation we perceive in using Excel is that Excel regression is limited to 16 independent variables. We can also use R for linear regression.

To install R, visit https://cran.rstudio.com/
Open a folder for R
Select Download R for windows

To install Rattle:
Open the R Desktop icon (32 bit or 64 bit) and enter the following command at the R prompt. R will ask for a CRAN mirror. Choose a nearby location.

> install.packages("rattle")

Enter the following two commands at the R prompt. This loads the Rattle package into the library and then starts up Rattle.

> library(rattle)
> rattle()

If the RGtk2 package has yet to be installed, there will be an error popup indicating that libatk-1.0-0.dll is missing from your computer. Click on the OK and then you will be asked if you would like to install GTK+. Click OK to do so. This then downloads and installs the appropriate GTK+ libraries for your computer. After this has finished, do exit from R and restart it so that it can find the newly installed libraries.

When running Rattle a number of other packages will be downloaded and installed as needed, with Rattle asking for the user's permission before doing so. They only need to be downloaded once.

Figure 3.7 shows an initial screen where we load the monthly data file.

We select *Spreadsheet*. The data file is linked in the *Filename* menu. When we click on *Execute*, we see the data types. Price was originally numeric, but we select *Target* as this is what we wish to predict. Again, click on *Execute* to induce R to read price as the target.

We now want to run a linear model. Click on the *Model* tab, yielding Fig. 3.8.

R displays its options under linear models. Using the monthly data, we select *Numeric* and *Linear*, and click on *Execute*. This yields the output shown in Table 3.12.

This model is not OLS, as R applies iterative fit. The model here is:

$$\text{Brent Price} = 35.05602 + 0.37295 \times \text{month}$$

This is different than the Excel OLS model, and not as strong a fit as the r-squared here is only 0.37, compared to 0.43 from the Excel OLS model. Table 3.13 shows the model from R's linear model for daily data.

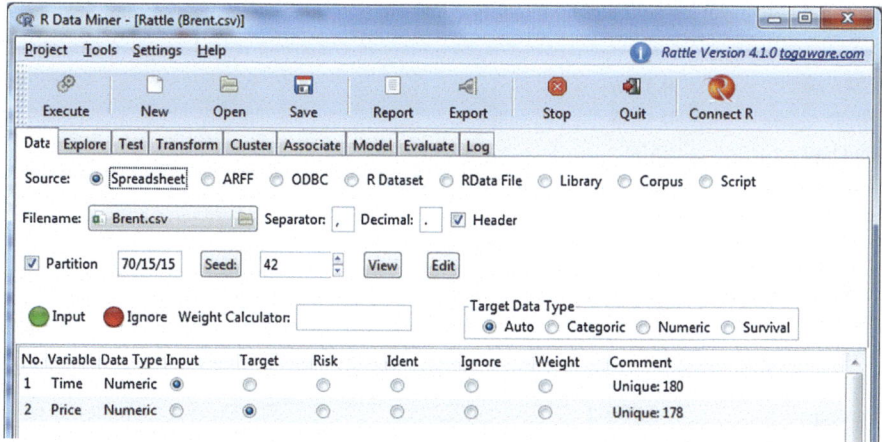

Fig. 3.7 Rattle data-loading screenshot

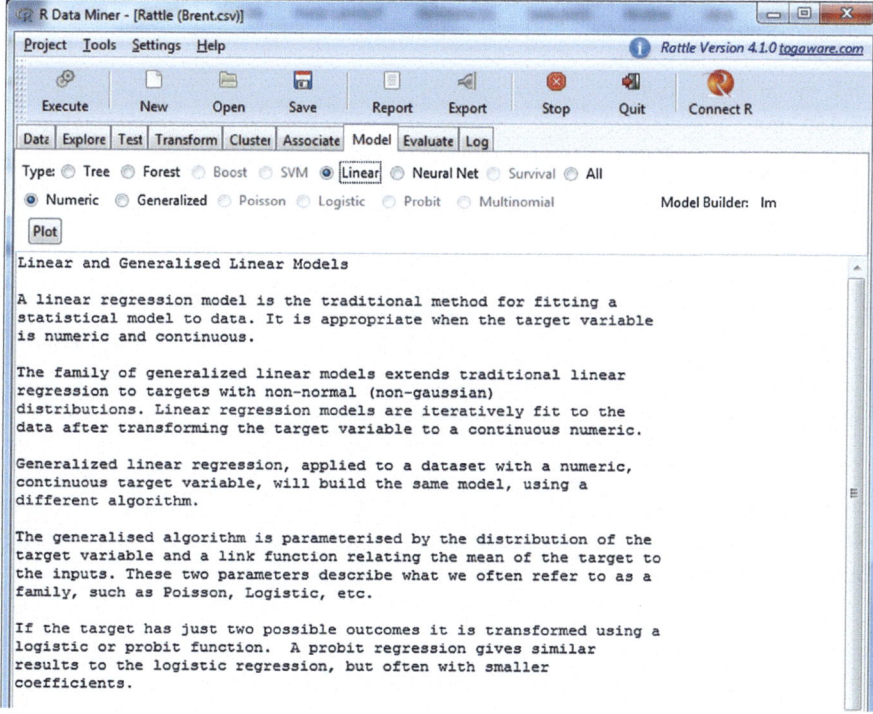

Fig. 3.8 Linear regression screenshot

Table 3.12 R linear model output—monthly crude oil

Regression statistics				
R square	**0.3706**			
Adj R square	0.3655			
Standard error	25.71			
Observations	124			
ANOVA				
	df	SS	MS	F
Regression	1	47882	47882	72.436
Residual	123	81305	661	
Total	124	129187		
	Coefficients	Standard error	t Stat	*P*-value
Intercept	**35.05602**	4.72275	7.423	1.65e-11
Time	**0.37295**	0.04382	8.511	4.99e-14

Table 3.13 R linear model output—daily crude oil

Regression statistics				
R square	**0.4676**			
Adj R square	0.4674			
Standard error	23.91			
Observations	2908			
ANOVA				
	df	SS	MS	F
Regression	1	1458773	1458773	2552.7
Residual	2907	1661222	571	
Total	2908	3119995		
	Coefficients	Standard error	t Stat	*P*-value
Intercept	**26.5831**	0.8833975	30.09	2e-16
Time	**0.0187**	0.0003693	50.52	2e-16

This model has a better fit than the monthly data, with an r-squared measure of 0.4676. This is in part due to the much larger sample size, a feature of data mining. The resulting formula:

$$\text{Brent price} = 26.5831 + 0.0187$$
$$\times \text{ time where time is day beginning 2 January 2001}$$

We also can regress against daily change in crude price, which Fig. 3.5 showed to be centered very close to zero (Table 3.14).

Note here the miniscule r-squared, as well as beta coefficients. The model and beta coefficients are only slightly significant (at the 0.05 level), which means that

Table 3.14 R lm model output for daily change in brent crude oil

Regression statistics				
R square	**0.001498**			
Adj R square	0.001154			
Standard error	0.02225			
Observations	2908			
ANOVA				
	df	SS	MS	F
Regression	1	0.00216	0.00215844	4.36
Residual	2907	1.43912	0.00049505	
Total	2908	1.44128		
	Coefficients	Standard error	t Stat	P-value
Intercept	**0.0019946127**	0.0008222264	2.426	0.0153
Time	**−0.0000007177**	0.0000003437	−2.088	0.0369

the intercept and trend are quite close to zero (and in fact the trend value is actually negative—barely). This implies that if an investor in crude oil knew today's price, it would be a shot in the dark on average about whether the price would go up or down tomorrow.

3.8 Summary

There are many ways to extend time series. Moving average is one of the easiest, but can't forecast very far into the future. Regression models have been widely used in classical modeling. They continue to be very useful in data mining environments, which differ primarily in the scale of observations and number of variables used. Classical regression (usually ordinary least squares) can be applied to continuous data. Regression can be applied by conventional software such as SAS, SPSS, or EXCEL. R provides a linear model akin (but slightly different from) to OLS. There are many other forecasting methodologies, to include exponential smoothing. We covered the methods that relate to the techniques we will cover in future chapters. We have also initially explored the Brent crude oil data, demonstrating simple linear models and their output.

Chapter 4
Multiple Regression

Regression models allow you to include as many independent variables as you want. In traditional regression analysis, there are good reasons to limit the number of variables. The spirit of exploratory data mining, however, encourages examining a large number of independent variables. Here we are presenting very small models for demonstration purposes. In data mining applications, the assumption is that you have very many observations, so that there is no technical limit on the number of independent variables.

4.1 Data Series

We will model the monthly price of the MSCI (Morgan Stanley Capital International) China Index. MSCI is intended to reflect Chinese mainland equity markets, to include trade on the Shanghai, Shenzhen, and Hong Kong exchanges, of both state-owned and non-state-owned shares. Monthly data for the period January 2001 through April 2016 yielded the time series regression shown in Table 4.1.

It can be seen from Table 4.1 that time has explained nearly 65 % of the change in the index value. However, the series is quite volatile. This regression model is plotted against the MSCI in Fig. 4.1, showing this volatility.

Multiple ordinary least squares (OLS) regression allows consideration of other variables that might explain changes in MSCI. We consider five additional variables, with the intent of demonstrating multiple regression, not with the aim of completely explaining change in MSCI. The variables we include the S&P 500 index of blue chip stocks in the US, the New York Stock Exchange (NYSE) index of all stocks traded on that exchange, both reflecting US capital performance, a possible surrogate for the US economy, which has been closely tied to the Chinese economy over the time period considered. Eurostoxx is an index of European

© Springer Science+Business Media Singapore 2017
D.L. Olson and D. Wu, *Predictive Data Mining Models*,
Computational Risk Management, DOI 10.1007/978-981-10-2543-3_4

Table 4.1 MSCI monthly time series regression

Summary output						
Regression statistics						
Multiple R	0.80357					
R square	**0.645724**					
Adj R square	0.643778					
Standard error	12.37509					
Observations	184					
ANOVA						
	df	*SS*	*MS*	*F*	Significance *F*	
Regression	1	50801.15	50801.15	331.7239	7.16E−43	
Residual	182	27,872	153.1429			
Total	183	78673.16				
	Coefficients	Standard Error	*t* Stat	*P*-value	Lower 95 %	Upper 95 %
Intercept	**17.68337**	1.83207	9.652122	4.63E−18	14.06854	21.29819
Time	**0.312829**	0.017176	18.21329	7.16E−43	0.27894	0.346718

Fig. 4.1 MSCI Time Series

stocks, another Chinese trade partner. Each of these three stock indices were obtained from http://finance.yahoo.com. Brent is the price of Brent crude oil, obtained from www.tradingeconomics.com/commodity/brent-crude-oil, reflecting a cost of doing business for Chinese industry as well as for the rest of the world. Brent crude oil price can be viewed as a reflection of risk, as was seen in Chap. 3. The last variable considered is the price of gold, which can be viewed as a measure of investor uncertainty (when things are viewed as really bad, investors often flee to gold). The price of gold was obtained from http://goldprice.org/gold-price–history/html. All data used was monthly for the period January 2001 through April 2016.

4.2 Correlation

Since we have multiple independent variable candidates, we need to consider first their strength of contribution to the dependent variable (MSCIchina) , and second, overlap in information content with other independent variables. We want high correlation between MSCIchina and the candidate independent variables. We want low correlation between independent variables. Table 4.2 provides correlations obtained from Excel.

The strongest relationship between MSCIchina and candidate independent variables is with Time, at 0.80357. You might note that this r when squared (0.645724) equals the r-squared output from Table 4.1 above. That is because the r in question (correlation coefficient) is in fact the same r that is squared in the single regression output. While the trend is strong (and the output from Table 4.1 in fact indicates by p-value that it is highly significant), it only explains about 65 % of the change in MSCIchina. The next independent variable by correlation is Gold (at 0.773), followed by NYSE (0.748), Brent (0.746), and S&Pclose (0.599). Eurostoxx have a very low correlation with MSCIchina, and can be disregarded.

We now need to worry about overlapping information content. Time has strong correlation with Gold, S&Pclose, and NYSE (and to a slightly lesser extent, with Brent). NYSE and S&Pclose have a very high correlation, not unexpectedly as they both measure US investor confidence. Because of this high correlation, we should only include one of them as independent variables in predicting MSCIchina. We choose NYSE due to its higher correlation with MSCIchina. Next we consider Gold and Brent. These both represent risk to some degree (Brent being a source, Gold being a response to perceived risk). The correlation between Gold and Brent is quite high at 0.82. Thus we only want to include one of this pair. Gold has higher overlap with time, so we select Brent, giving us three independent variables in a multiple regression model MSCIchina = f{Time, NYSE, Brent}. The Excel output for this model is given in Table 4.3.

The mathematical model is thus:

$$\text{MSCIchina} = -9.62453 + 0.135166 \times \text{Time} + 0.003577 \times \text{NYSE} + 0.228168 \times \text{Brent}$$

Table 4.2 Correlations among variables

	Time	MSCIchina	S&Pclose	NYSE	Eurostoxx	Brent	Gold
Time	1						
MSCIchina	**0.80357**	1					
S&Pclose	0.741268	0.598793	1				
NYSE	**0.737029**	**0.748186**	0.940107	1			
Eurostoxx	−0.21211	0.118288	0.331011	0.424784	1		
Brent	**0.657869**	**0.746425**	0.350836	**0.536606**	−0.13858	1	
Gold	0.88539	0.773361	0.48211	0.527368	−0.35537	0.818475	1

Table 4.3 Multiple regression OLS prediction model for MSCIchina

Summary output						
Regression statistics						
Multiple R	0.877666					
R square	0.770297					
Adj R square	**0.766468**					
Standard error	10.01984					
Observations	184					
ANOVA						
	df	*SS*	*MS*	*F*	Significance *F*	
Regression	3	60601.68	20200.56	201.2066	3.02E−57	
Residual	180	18071.48	100.3971			
Total	183	78673.16				
	Coefficients	Standard Error	*t* Stat	*P*-value	Lower 95 %	Upper 95 %
Intercept	**−9.62453**	3.892611	−2.47251	0.014345	−17.3056	−1.94351
Time	**0.135166**	0.023175	5.83247	2.48E−08	0.089437	0.180895
NYSE	**0.003577**	0.000631	5.66506	5.73E−08	0.002331	0.004823
Brent	**0.228168**	0.030516	7.477004	3.21E−12	0.167953	0.288383

The *p*-value for the intercept doesn't really matter as we need an intercept in most cases. The *p*-values for the other three variables are all highly significant. Plotting this model versus actual MSCI and its trend line is shown in Fig. 4.2.

Clearly the multiple regression fits the data better. This is indicated quantitatively in the r-square (0.770) which has to be greater than the 0.646 r-squared of the simple regression against Time, because Time is included in the multiple regression. Adding independent variables will always increase r-squared. To get a truer

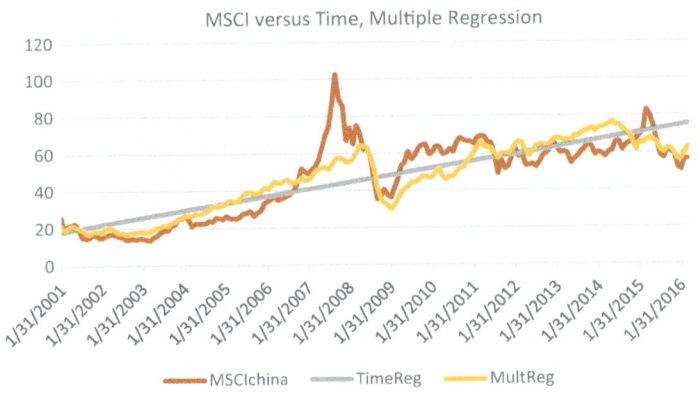

Fig. 4.2 Multiple Regression versus MSCIchina actual and trendline

picture of the worth of adding independent variables to the model, adjusted R^2 penalizes the R^2 calculation for having extra independent variables.

$$\text{Adjusted}\,R^2 = 1 - \frac{\text{SSE}(n-1)}{\text{TSS}(n-k)}$$

where

SSE	sum of squared errors
MSR	sum of squared predicted values
TSS	SSE + MSR
n	number of observations
k	number of independent variables.

Adjusted R^2 can be used to select more robust regression models. We might get almost as much predictive accuracy without needing to gather these variables. Here Adj R Square goes up from 0.644 to 0.766, indicating that adding the two additional independent variables paid their way in additional explanatory power.

Using the model to forecast requires knowing (or guessing at) future independent variable values. A very good feature for Time is that there is no additional error introduced in estimating future time values. That is not the case for NYSE or for Brent. In Table 4.4 we guess at slight increases for both NYSE and Brent values, and compare the simple time regression with the multiple regression model.

4.3   Lags

One way to avoid introducing extra error by guessing at future independent variable values is to use lags. This regresses the dependent variable against the independent variable values lagged by one or more periods. There usually is a loss of model fit, but at least there is no extra error introduced in the forecast from guessing at the independent variable values. Here we show the results of lagging one, two, and three periods into the future. Table 4.5 shows the model for a lag of one period for NYSE and Brent.

Table 4.4 Forecasts

Time	NYSE	Brent	TimeReg	MultReg
185	10,500	50	75.56	64.35
186	10,600	51	75.87	65.07
187	10,700	52	76.18	65.79
188	10,800	53	76.50	66.51
189	10,900	54	76.81	67.24
190	11,000	55	77.12	67.96

Table 4.5 Lag-1 model

Summary output						
Regression statistics						
Multiple R	0.864368					
R square	0.747132					
Adj R square	0.742894					
Standard error	10.51388					
Observations	183					
ANOVA						
	df	SS	MS	F	Significance F	
Regression	3	58463.25	19487.75	176.2934	3.36E-53	
Residual	179	19786.94	110.5416			
Total	182	78250.2				
	Coefficients	Standard Error	t Stat	P-value	Lower 95 %	Upper 95 %
Intercept	−7.49978	4.074192	−1.8408	0.067306	−15.5394	0.539843
Time	0.149662	0.024589	6.086613	6.84E−09	0.101141	0.198183
NYSE-1	0.003351	0.000663	5.052894	1.07E−06	0.002042	0.00466
Brent-1	0.204356	0.032573	6.273692	2.58E−09	0.140079	0.268633

To predict with this model, we know the next month Time = 185, and we now have values for NYSE one period prior (10436.92) and Brent one period prior (45.83). This yields a prediction of:

$$
\begin{aligned}
MSCIchina\,(185) &= -7.49978 + 0.149662 \times 185 + 0.003351 \\
&\quad \times 10436.92 + 0.204356 \times 45.83 \\
&= 66.174
\end{aligned}
$$

We present results of lags for two additional periods in Table 4.6.

It can be seen from Table 4.6 that the fit of the regression declines with longer lags.

We can demonstrate the value of adjusted r-squared by looking at the R linear model output for a model including Time as well as the price of gold, the price of oil, and the three other stock indices we have presented. This yields the R output in Fig. 4.3.

Table 4.6 Lagged models

Time	Concurrent model	Lag 1	Lag 2	Lag 3
185		66.17	63.94	62.70
186			66.01	65.58
187				67.41
r-squared	0.770	0.747	0.722	0.699
Adj r-squared	0.766	0.743	0.717	0.693

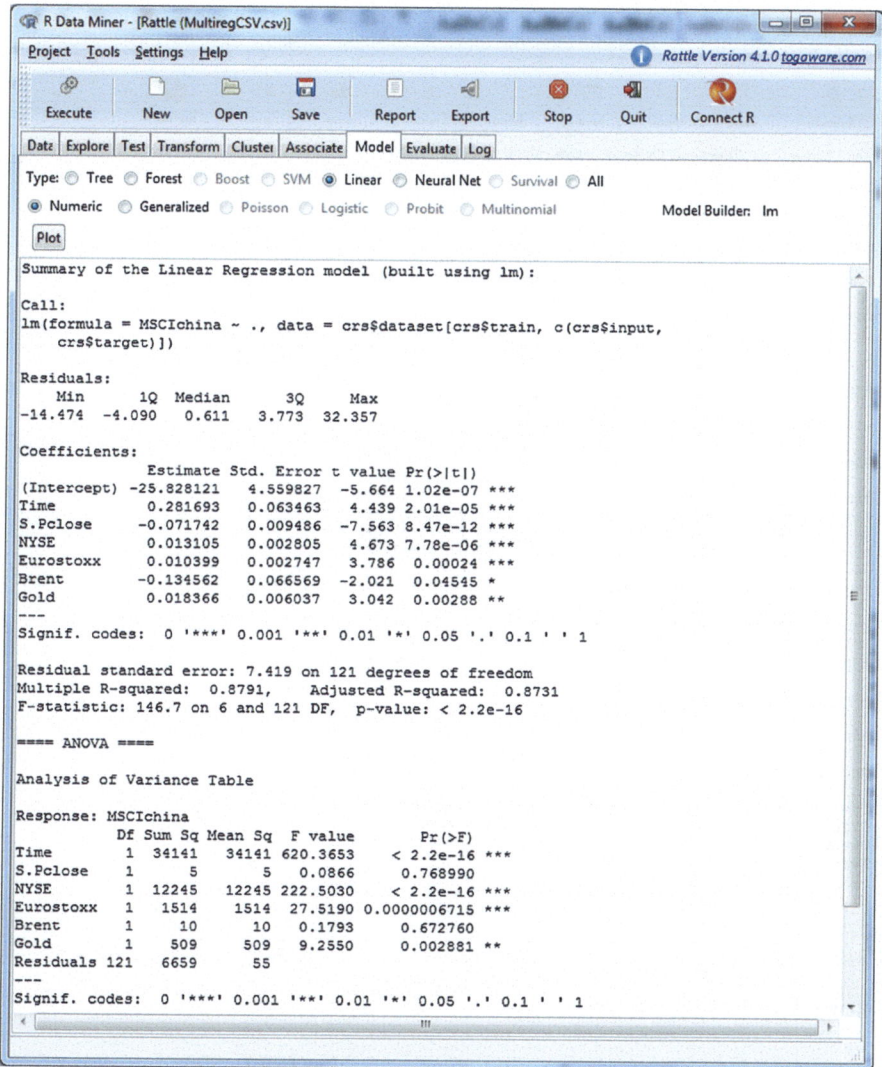

Fig. 4.3 R output for multiple regression with all variables

This output shows an r-squared value of 0.8791 (adjusted r-squared 0.8731). Two of the beta coefficients are marginally significant (Brent has almost a 0.05 probability of being insignificant; gold a safer 0.003 probability). But as we have seen, there is a high probability of multicollinearity (overlapping information content among independent variables). Thus we run a trimmed model using Time, NYSE, and Brent as independent variables, yielding the output in Fig. 4.4.

Here we can see that r-squared declines, as it must when we eliminate independent variables. Adjusted r-squared also drops, from 0.8731 to 0.7428. Thus if

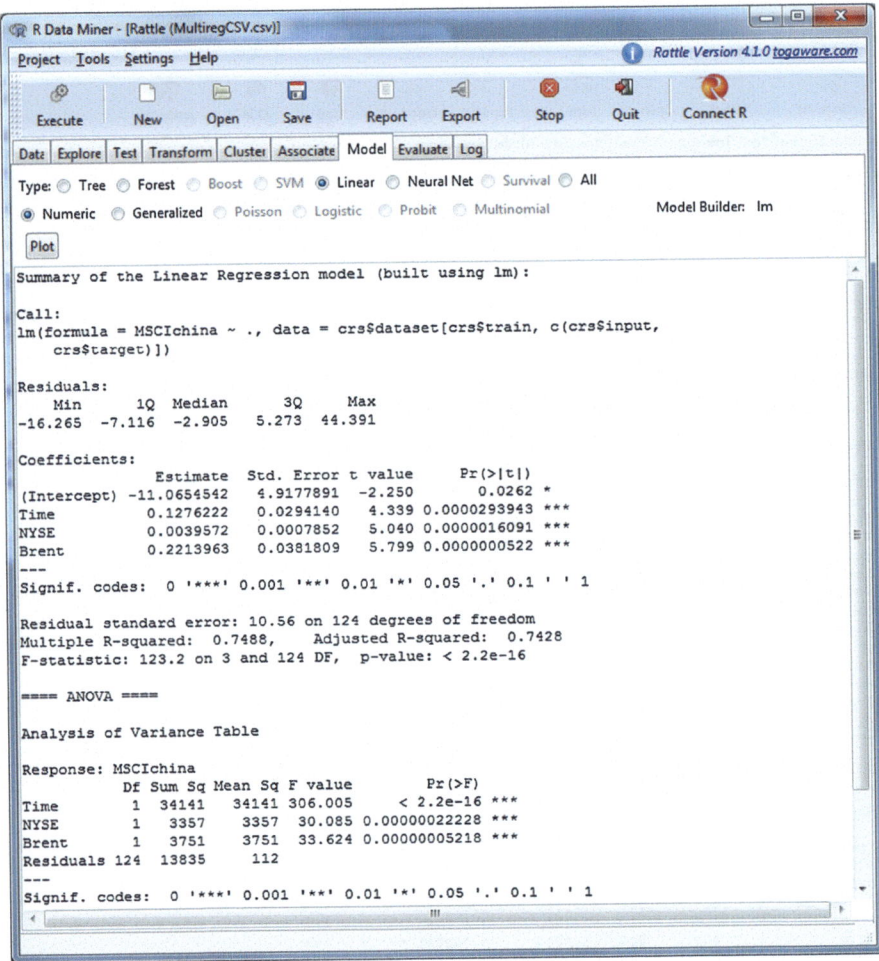

Fig. 4.4 R multiple trimmed regression

we had valid forecasts for the eliminated variables (S&P 500, Eurostoxx, and Gold), we should keep the bigger model from Fig. 4.3. However, given their uncertainty, it is sounder to use this trimmed model. Table 4.7 displays a possible forecast using the model.

$$\text{MSCIchina} = -11.0655 + 0.1276 \times \text{Time} + 0.0040 \times \text{NYSE} + 0.2214 \times \text{Brent}$$

Time	NYSE	Brent	Forecast MSCIchina
185	10,500	50	65.6
186	10,600	55	67.2
187	10,700	60	68.9

Table 4.7 Forecast from Fig. 4.4

Table 4.8 Alternative
forecast from Fig. 4.4

Time	NYSE	Brent	Forecast MSCIchina
185	10,000	49	63.4
186	9500	47	61.1
187	10,500	45	64.8

Note that we rounded the beta coefficients as shown in our formula. This led to variance of almost one index point for MSCIchina. But given the error introduced by our estimates for NYSE and Brent, this is a minor variance. If we had been less optimistic, as in Table 4.8, the forecast would have been quite different.

We hope we have demonstrated that adding more variables to a model gives an apparent improvement in forecasting, yet it introduces a hidden error in the estimates of future independent variable values.

4.4 Summary

Regression models have been widely used in classical modeling. They continue to be very useful in data mining environments, which differ primarily in the scale of observations and number of variables used. Classical regression (usually ordinary least squares) can be applied to continuous data. Regression can be applied by conventional software such as SAS, SPSS, or EXCEL. The more independent variables, the better the fit. But for forecasting, the more independent variables, the more things that need to be identified (guessed at), adding to unmeasured error in the forecast.

There are many refinements to regression that can be accessed, such as stepwise linear regression. Stepwise regression uses partial correlations to select entering independent variables iteratively, providing some degree of automatic machine development of a regression model. Stepwise regression has its proponents and opponents, but is a form of machine learning.

Chapter 5
Regression Tree Models

Decision trees are models that process data to split it in strategic places to divide the data into groups with high probabilities of one outcome or another. It is especially effective at data with categorical outcomes, but can also be applied to continuous data, such as the time series we have been considering. Decision trees consist of nodes, or splits in the data defined as particular cutoffs for a particular independent variable, and leaves, which are the outcome. For categorical data, the outcome is a class. For continuous data, the outcome is a continuous number, usually some average measure of the dependent variable.

Witten and Frank [1] describe the use of decision trees for numeric prediction as regression trees, based upon statistical use of the term regression for the operation of computing numeric quantities with averaged numeric values. Regression equations can be combined with regression trees, as in the M5P model we will present below. But a more basic model is demonstrated with R, here used to predict MSCIchina considering all of the candidate independent variables in our dataset.

5.1 R Regression Trees

The variables in the full data set are: {Time, S&P500, NYSE, Eurostoxx, Brent, and Gold}. Opening R and loading the dataset MultiregCSV.csv as we have in Chap. 4, executing, and selecting the "Model" tab, we obtain the screen in Fig. 5.1.

This screen informs us that R is applying a model similar to CART or ID3/C4, widely known decision tree algorithms. Executing this in R yields the output in Fig. 5.2.

The decision tree provided by Fig. 5.2 output is displayed in Table 5.1.

The decision tree model in R selected 70 % of the input data to build its model, based on the conventional practice to withhold a portion of the input data to test (or refine) the model. Thus it used 128 of the 184 available observations. The input data values for MSCIchina ranged from 13.72 to 102.98 (in late 2007), with 2016

© Springer Science+Business Media Singapore 2017
D.L. Olson and D. Wu, *Predictive Data Mining Models*,
Computational Risk Management, DOI 10.1007/978-981-10-2543-3_5

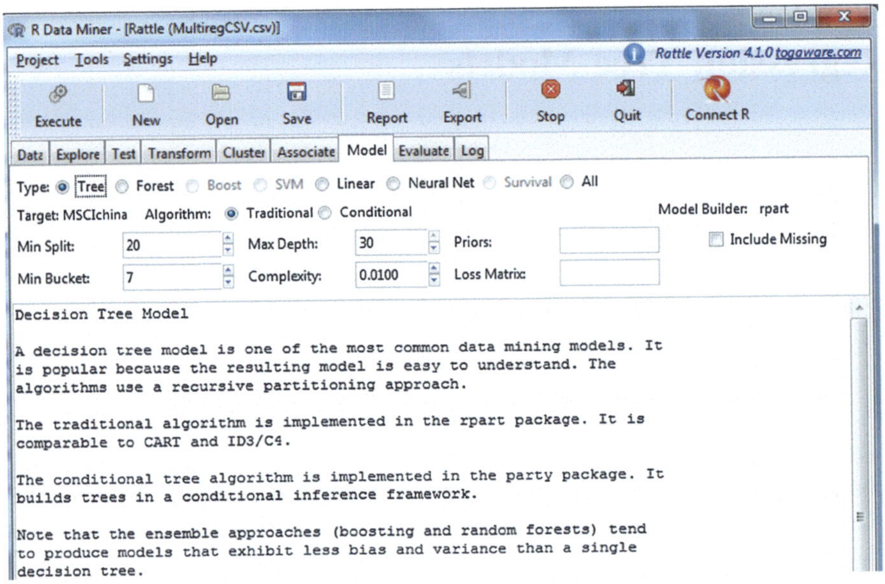

Fig. 5.1 R regression tree screen

monthly values in the 51–57 range. For forecasting, the first three rules don't apply (they are for the range January 2001 through October 2006). The last four rules thus rely on estimates of Eurostoxx (in the 3000s for the past year or so) and NYSE (which recently has been above and below the cutoff of 10,781). Thus this model realistically forecasts MSCIchina to be in the 60 s, reasonable enough given its values in the past few years.

We can also run this model with the trimmed data (not including Eurostoxx). Table 5.2 shows the R output, again based on 128 observations.

The only rules here that apply to forecasting are the fourth row where NYSE < 8047.2 (hasn't happened since August 2012) and the last row, which forecasts an MSCIchina value of 64.04. This is obviously far more precise than is realistic, but falls within the range of the model using all variables.

5.2   WEKA Regression Trees

WEKA provides an open source software that can be downloaded from www.cs. waikato.ac.nz/ml/weka/. The download comes with documentation.

Upon opening WEKA, Hit the Open file … button on the upper left.

Link to LoanRaw.csv (or any.csv or .arff file you want to analyze).

Install, obtaining Fig. 5.3.

Select *Explorer*, yielding Fig. 5.4.

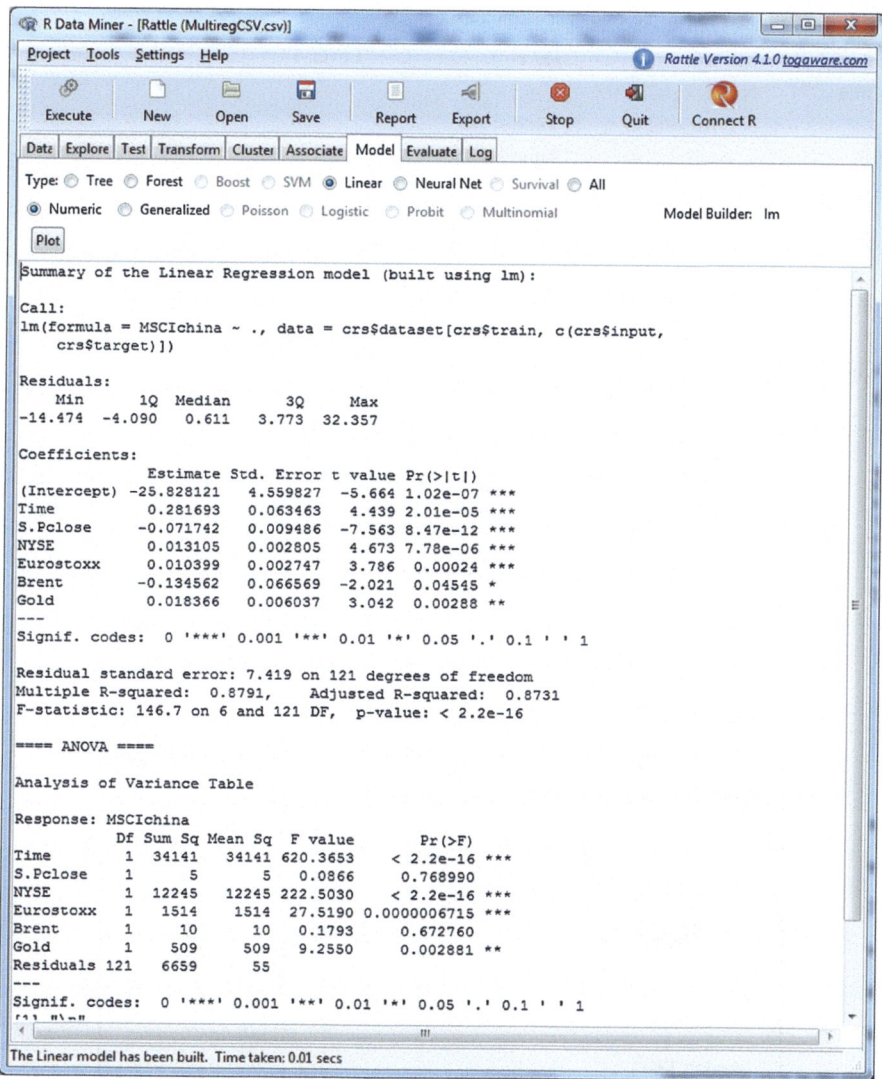

Fig. 5.2 R regression tree output for full dataset

You can then link the data by selecting the "Open file.." button, and select
MultiregCSV.cxv data (WEKA won't read Excel files directly, but will read comma
separated variable data). The Classify tab has two useful regression trees—M5P and
RepTree.

The *Choose* tab in Fig. 5.5 allows accessing a number of models within WEKA.
Because our data is numeric, only DecisionStump, M5P, and RepTree are available.
DecisionStump is a simplistic algorithm, so we will look at the other two. The M5P
tree learner builds logistic model trees.

Table 5.1 R regression tree rules for full dataset

Level 1	Level 2	Level 3	Level 4	Cases	Forecast
Time < 71	NYSE < 6286.09			18	15.85
	NYSE > 6286	Time < 60		19	25.42
		Time ≥ 60		9	36.43
Time ≥ 71	Eurostoxx < 2465			11	50.93
	Eurostoxx ≥ 2465	Eurostoxx < 4238	NYSE < 10,781	52	61.78
			NYSE ≥ 10,781	11	69.95
		Eurostoxx ≥ 4238		8	74.28

Table 5.2 R regression tree for trimmed data

Level 1	Level 2	Level 3	Level 4	Cases	Forecast
Time < 71	NYSE < 6286.09			18	15.85
	NYSE > 6286	Time < 60		19	25.42
		Time ≥ 60		9	36.43
Time ≥ 71	NYSE < 8047.2			24	56.78
	NYSE ≥ 8047.2	Time < 80		7	56.66
		Time ≥ 80	Time < 89	7	79.86
			Time ≥ 89	44	64.04

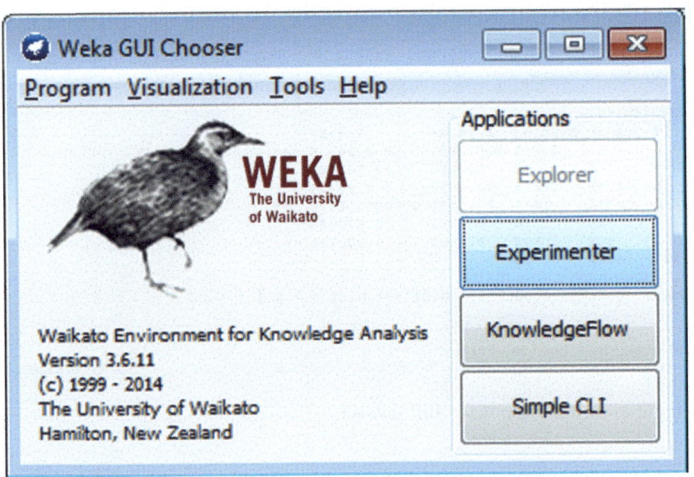

Fig. 5.3 WEKA opening screen

5.2.1 M5P Modeling

M5P uses cross-validation to determine how many iterations to run rather than cross-validating at every node. The minimum number of cases per rule can be set

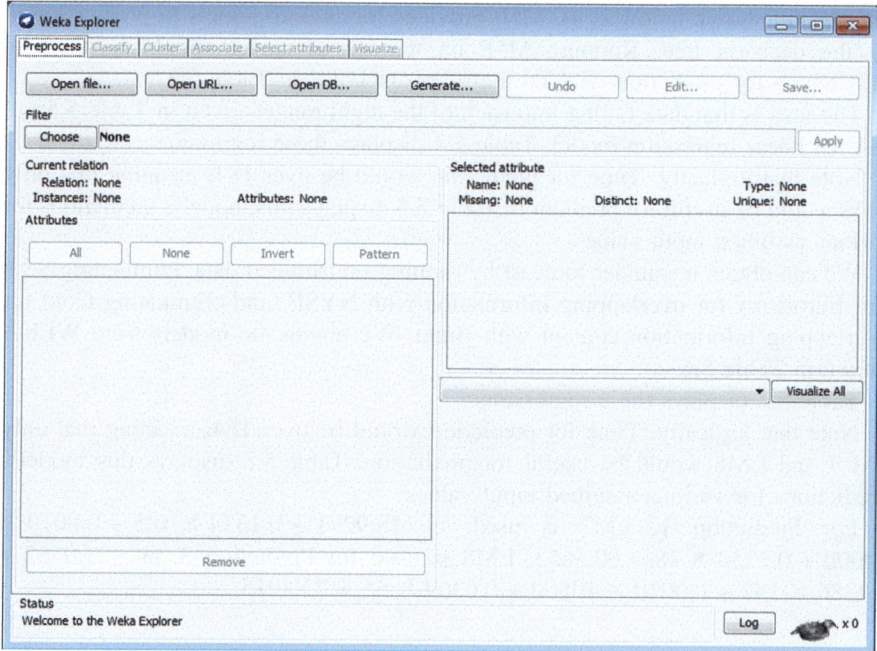

Fig. 5.4 WEKA explorer screen

Fig. 5.5 WEKA Classify Screen

(the default shown below is 4). M5P provides regression models to predict sorted by the decision tree. Running M5P on the set of seven variables, selecting MSCIchina for prediction, yielded the rules displayed in Table 5.3.

The idea is that data falling into each of the eight ranges given in Table 5.5 has its own linear regression model. Table 5.4 displays these regressions.

Note that logically, Time for prediction would be over 184, meaning that only LM8 would be useful for prediction. Table 5.5 displays this model's predictions for various assumed input values.

We can obtain a sounder forecast by running on trimmed data, eliminating S&P and Eurostoxx for overlapping information with NYSE, and eliminating Gold for overlapping information content with Brent. We obtain the models from WEKA shown in Table 5.6.

Table 5.7 displays these regressions.

Note that logically, Time for prediction would be over 184, meaning that only LM 7 and LM8 would be useful for prediction. Table 5.8 displays this model's predictions for various assumed input values.

For Prediction 1, LM7 is used, or $18.9271 + 0.1374 \times 185 + 0.0010 \times 10000 + 0.1254 \times 48 = 60.3653$. LM8 is used for Prediction 3, or $-21.0562 + 0.4386 \times 187 + 0.0010 \times 10500 + 0.0369 \times 55 = 73.4915$.

5.2.2   REP Tree Modeling

WEKA also has a REPTree algorithm, which builds the regression tree using gain/variance reduction with pruning based on error reduction. This yielded a complex tree for the trimmed data with 24 rules involving up to 11 levels, but we can simplify by only looking at rules where Time was over 184 in Table 5.9.

Given we want to predict for Time > 184, this model basically says that if NYSE prediction is < 5970, it forecasts MSCIchina to be 42.01, if NYSE is predicted to be over 10,715, it predicts MSCIchina to be 76.10, and if NYSE is between those limits, the forecast for MSCIchina is 54.03. Given our assumptions from Table 5.8 for NYSE, the prediction of the REPTree model is 54.03.

5.3   Random Forests

R has a random forest modeling capability, accessed as shown in Fig. 5.6.

Note that we don't have a particularly large data set, so a random forest is not particularly appropriate for this data.

Table 5.3 M5P rules—full training set

Level 1	Level 2	Level 3	Level 4	Level 5	Level 6	Linear model
Time ≤ 69						LM1
Time > 69	Time ≤ 100					LM2
	Time > 100	Eurostoxx ≤ 2616.6	Eurostoxx ≤ 2465.8	Gold ≤ 1645.8		LM3
				Gold > 1645.8	NYSE ≤ 7506.4	LM4
					NYSE > 7506.4	LM5
			Eurostoxx > 2465.8			LM6
		Eurostoxx > 2616.6	Time ≤ 169			LM7
			Time > 169			LM8

Table 5.4 M5P linear models

Model	Intercept	Time	S&P	NYSE	Eurostoxx	Brent	Gold
LM1	−23.2927	0.0506	0.0091	0.0022	0.0019	−0.0169	0.0367
LM2	−142.1569	1.0264	−0.0290	0.0029	0.0283	−0.0109	0.0327
LM3	35.7252	0.0109	−0.0070	0.0021	0.0045	−0.0109	0.0022
LM4	33.0972	0.0109	−0.0070	0.0025	0.0045	−0.0109	0.0022
LM5	34.5265	0.0109	−0.0070	0.0024	0.0041	−0.0109	0.0022
LM6	35.3249	−0.0136	−0.0070	0.0021	0.0056	−0.0109	0.0034
LM7	13.1618	0.0109	−0.0076	0.0006	0.0130	−0.0405	0.0150
LM8	**−8.7343**	**0.0109**	**−0.0325**	**0.0090**	**0.0097**	**0.0021**	**0.0087**

Table 5.5 LM8 predictions

Independent variable	Prediction 1	Prediction 2	Prediction 3
Time	185	186	187
S&P	2100	2120	2140
NYSE	10,000	10,100	10,500
Eurostoxx	2800	2600	2300
Brent	48	50	55
Gold	1300	1200	1250
PREDICTION	**53.60**	**51.1**	**51.6**

Table 5.6 M5P rules —trimmed training set

Level 1	Level 2	Level 3	Level 4	Linear Model
Time < 69.5	NYSE < 6375.5			LM1
	NYSE > 6375.5	Time < 60.5		LM2
		Time > 60.5		LM3
Time > 69.5	Time < 100.5	Brent < 67.4		LM4
		Brent > 67.4		LM5
	Time > 100.5	NYSE < 10494.4	Time < 127.5	LM6
			Time > 127.5	LM7
		NYSE > 10494.4		LM8

Table 5.7 M5P linear models for trimmed model

Model	Intercept	Time	NYSE	Brent
LM1	−9.6630	0.1275	0.0040	0.0610
LM2	−11.9096	0.0536	0.0049	0.0426
LM3	−25.2457	0.0516	0.0069	0.0295
LM4	35.4535	−0.1876	0.0016	0.3363
LM5	109.0079	−0.9741	0.0016	0.2720
LM6	27.3869	0.0846	0.0032	−0.0060
LM7	**18.9271**	**0.1374**	**0.0010**	**0.1254**
LM8	**−21.0562**	**0.4386**	**0.0010**	**0.0369**

Table 5.8 Trimmed model M5P predictions

Independent variable	Prediction 1	Prediction 2	Prediction 3
Time	185	186	187
NYSE	10,000	10,100	10,500
Brent	48	50	55
Prediction	**60.4**	**60.9**	**73.5**

Table 5.9 REPTree model for time > 184

Level 1	Level 2	Level 3	Level 4	Level 5	Prediction
Time > 71	NYSE < 5969.82				42.01
	NYSE ≥ 5968.82	Time > 86	NYSE < 10714.68	Time > 180	54.03
			NYSE ≥ 10714.68	Time > 171	76.10

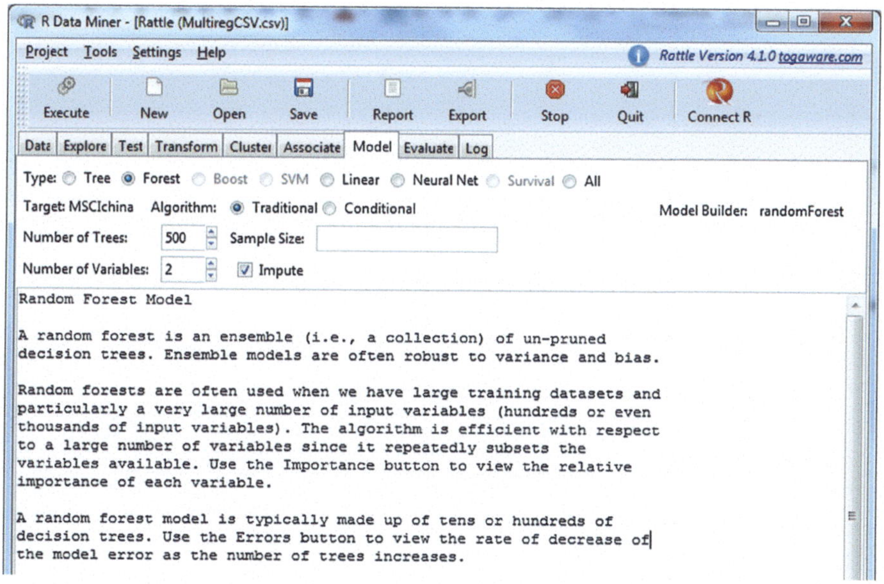

Fig. 5.6 R forest option

5.4 Summary

Regression trees offer an additional means to predict continuous data. You have seen the algorithms available from R and WEKA, and see that they are conditional regressions, with the decision tree providing a framework of conditions. As prediction tools, they are subject to the same limitations found with multiple regression—they include measurable error from past fit, but no way to measure the additional error introduced from the errors involved in guessing future independent variable values. Nonetheless, regression trees offer an additional tool available in dealing with predicting time series data.

Reference

1. Witten IH, Frank E (2005) Data mining: practical machine learning tools and techniques, 2nd edn. Elsevier, Amsterdam

Chapter 6
Autoregressive Models

Autoregressive models take advantage of the correlation between errors across time periods. Basic linear regression views this autocorrelation as a negative statistical property, a bias in error terms. Such bias often arises in cyclical data, where if the stock market price was high yesterday, it likely will be high today, as opposed to a random walk kind of characteristic where knowing the error of the last forecast should say nothing about the next error. Traditional regression analysis sought to wash out the bias from autocorrelation. Autoregressive models, to the contrary, seek to utilize this information to make better forecasts. It doesn't always work, but if there are high degrees of autocorrelation, autoregressive models can provide better forecasts.

There are two primary autoregressive models we will discuss. Autoregressive integrated moving average (ARIMA) models consider autocorrelation, trend, and moving average terms. Generalized autoregressive conditional heteroscedasticity (GARCH) models reduces the number of parameters required, and usually is good at capturing thick tailed returns and volatility clustering.

6.1 ARIMA Models

ARIMA models were designed for time series with no trend, constant variability, and stable correlations over time. ARIMA models have a great deal of flexibility. You must specify three terms:

1. **P**—the number of autocorrelation terms
2. **D**—the number of differencing elements
3. **Q**—the number of moving average terms

The **P** term is what makes a ARIMA model work, taking advantage of the existence of strong autocorrelation in the regression model Y = f(time).

© Springer Science+Business Media Singapore 2017
D.L. Olson and D. Wu, *Predictive Data Mining Models*,
Computational Risk Management, DOI 10.1007/978-981-10-2543-3_6

The **D** term can sometimes be used to eliminate trends. **D** of 1 will work well if your data has a constant trend (it's linear). **D** of 2 or 3 might help if you have more complex trends. Going beyond a **D** value of 3 is beyond the scope of this course. If there is no trend to begin with, **D** of 0 works well.

The model should also have constant variability. If there are regular cycles in the data, moving average terms equal to the number of observations in the cycle can eliminate these. Looking at a plot of the data is the best way to detect cyclical data. One easily recognized cycle is seasonal data. If you have monthly data, a moving average term **Q** of 12 would be in order. If you have quarterly data, **Q** of 4 should help. If there is no regular pattern, **Q** of 0 will probably be as good as any.

D and **Q** terms are used primarily to stabilize the data. **P** is the term which takes advantage of autocorrelation. The precise number of appropriate autocorrelation terms (**P**) to use can be obtained from the computer package. A correlogram provides autocorrelation coefficients, giving you an idea of the correlation of your dependent variable with lags of the same data. This correlogram will settle down to 0 (hopefully). **P** is the number of terms significantly different from 0. Significance is a matter of judgement. Since ARIMA models are often exploratory, you will want to try more than one model anyway, to seek the best fit (lowest mean square forecasting error). Stationarity tests diagnose the appropriate **D** terms called for.

ARIMA models tend to be volatile. They are designed for data sets of at least 100 observations. You won't always have that many observations. Both ARIMA and GARCH work better with larger data sets (although they can be applied to smaller ones). We will thus use Brent crude oil daily data in this chapter. We are looking at them as an alternative to time series, especially when autocorrelation is present in a regression versus time. So one idea is to compare different models and select the best one. Another approach that can be used in MATLAB is to use Bayesian information criterion (BIC) to select from a finite number of models.

6.1.1 ARIMA Model of Brent Crude

Matlab code yielding a time series forecast "r" for a given time series "x" is:

$$R = \text{price2ret}(x);$$

We first test a stationarity test. There are a number of such tests, including the Augmented Dickey-Fuller Test, the Kwiatkowski-Phillips-Schmidt-Shin Test, and the Phillips-Perron Test available at Matlab software. Using a 5 % significance level on the Brent crude data from January 2001 through December 2015, the Augmented Dickey-Fuller test is called as follows:

$$[\text{h,pValue}] = \text{adftest}(r, \text{`model'}, \text{`TS'}, \text{`lags'}, 0 : 5)$$

Here we obtain h values of 1 for all six lags.

The Kwiatkowski-Phillips-Schmidt-Shin test is called as follows:

$$[\text{h,pValue}] = \text{kpsstest}(r, \text{`Lags'}; , 1:5, \text{`Trend'}, \text{true})$$

The results are 0 for all five lags.
The Phillips-Perron test is called as:

$$[\text{h,pValue}] = \text{pptest}(r, \text{`model'}, \text{`TS'}, \text{`lags'}, 0:5)$$

The results are 1 for all lags.

These computations indicate that the return series is stationary at the 0.05 significance level for lags 1 through 5. The conclusion is the **D** is not needed in an ARIMA model for this data. To help determine **p** values, we can plot curves to evaluate series stationarity in Fig. 6.1.

Figure 6.1 demonstrates that lags of 0, 1, and 2 might be of interest. The ARIMA model is called by arima(p,D,q) to represent the general formula:

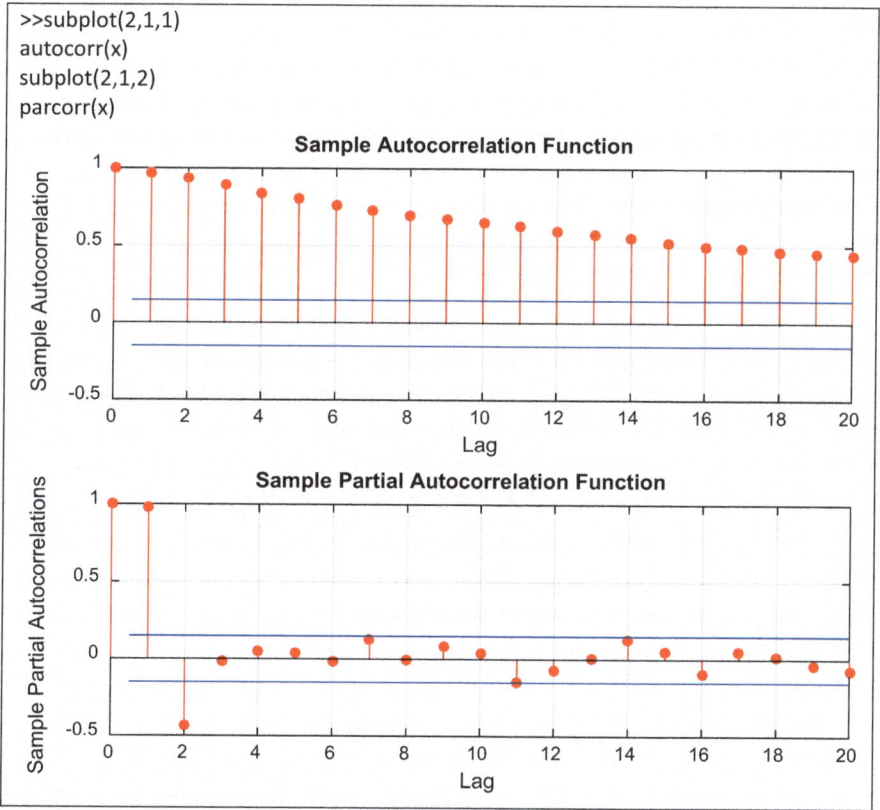

Fig. 6.1 Matlab autocorrelation diagnostics

$$\Delta^{D} y_t = c + \phi_1 \Delta^{D} y_{t-1} + \cdots + \phi_p \Delta^{D} y_{t-p} + \epsilon_t + \theta_1 \epsilon_{t-1} + \cdots + \theta_q \epsilon_{t-q}$$

where $\Delta^{D} y_t$ is a Dth differenced time series. You can write this model in condensed form using lag operator notation:

$$\phi(L)(1-L)^{D} y_t = c + \theta(L)\varepsilon_t$$

By default, all parameters in the model have unknown values, and the innovation distribution is Gaussian with constant variance. Several software products such as Stata or EVIEWS include pre-packaged commands for common tasks. Matlab is more flexible but thus more involved. The function arima() creates model objects for stationary or unit root nonstationary linear time series models, to include moving average (MA), autoregressive (AR), mixed autoregressive and moving average (ARMA), integrated (ARIMA), multiplicative seasonal, and linear time series models including a regression component (ARIMAX). Since we are using a **D** parameter value of 0, we have an ARMA model.

6.1.2 ARMA

For ARMA(p,q), we first need decide on the order, i.e., the value of **p** and **q** in ARMA. This can be done by depicting autocorrelation function (ACF) and partial autocorrelation function (PACF).

By observing ACF and PACF plots from Fig. 6.2, we approximately see that the values of p and q are 1 to 6 respectively. We further judge this by BIC rule, when we pick 1 to 6 for both **p** and **q** to compute the BIC value. Model selection statistics such as the Akaike information criterion (AIC) and Bayesian information criterion (BIC) to decide on the function order. When fitting models, adding parameters is likely to give a better fit over training data, but risks overfitting. AIC and BIC provide metrics to evaluate whether adding parameters is likely to provide a stabler model. We will use BIC, where lower (or more negative) measures are preferred. Figure 6.3 shows the Matlab code to call BIC for these combinations:

As can be seen from the above panel, BIC value for ARMA(1,1) of -353.3395 is the minimum, indicating that ARMA(1,1) or ARIMA(1,0,1) is the best choice. Figure 6.4 shows the Matlab call for this model, with results.

This Matlab input yields the diagnostics in Fig. 6.5.

Using these parameters in the model we obtain:

$$r_t = 0.00197161 + 0.318231 r_{t-1} + \epsilon_t - 0.0222004\epsilon_{t-1}$$

$$VAR(\epsilon_{t-1}) = 0.00752877$$

With the calibrated model, we can depict conditional variance, standardized residual, sample autocorrelation, and sample partial autocorrelation, which are used

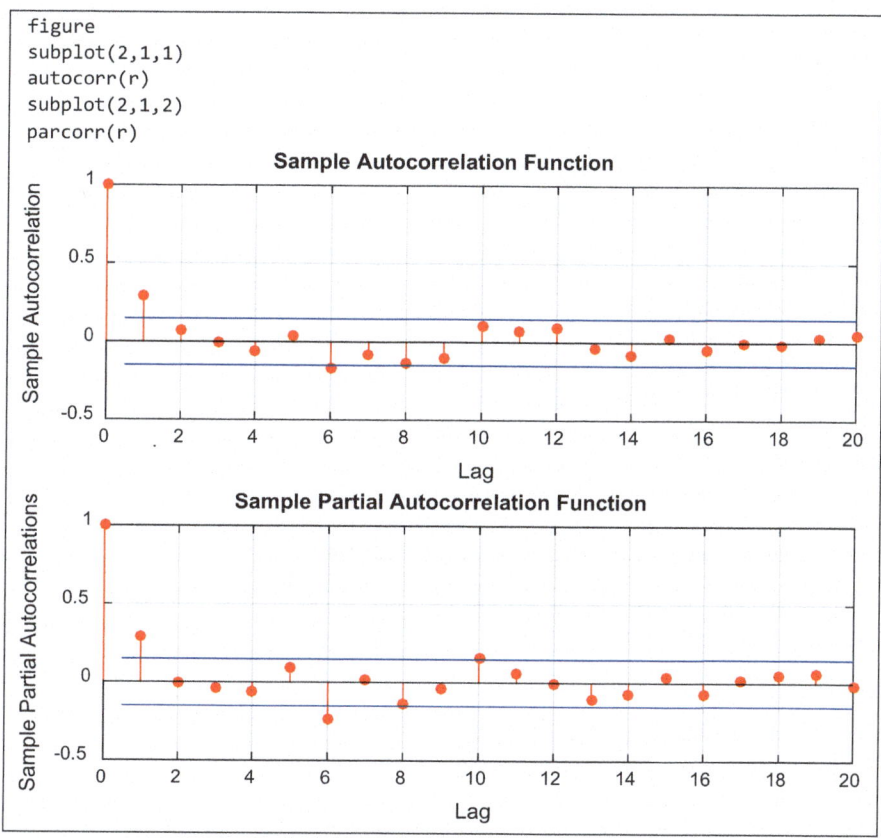

Fig. 6.2 ACF and PACF for daily brent crude oil

to test model validity. The plots indicate that the residual distribution is not very skewed. Both the residual ACF and PACF are small, suggesting a low autocorrelation. In other words, the model fits the data well.

We can use this model to predict. The Matlab code in Fig. 6.6 yields predictions for 50 days into the future.

Figure 6.7 displays these predictions graphically.

Table 6.1 gives the numeric forecasts for the first ten days.

To demonstrate the detailed calculation, the predicted mean proportional return on the first day is

$$r_{T+1} = 0.00197161 + 0.318231 r_T (=0.1628) + \epsilon_{T+1} (=0)$$
$$-0.0222004 \epsilon_T (=0.1399)$$
$$= 0.05067.$$

```
>> LOGL = zeros(6,6); %Initialize
PQ = zeros(6,6);
for p = 1:6
  for q = 1:6
    mod = arima(p,0,q);
    [fit,~,logL] = estimate(mod,r,'print',false);
    LOGL(p,q) = logL;
    PQ(p,q) = p+q;
  end
end
LOGL = reshape(LOGL,36,1);
PQ = reshape(PQ,36,1);
[~,bic] = aicbic(LOGL,PQ+1,100);
reshape(bic,6,6)

ans =

  -353.3395 -348.7345 -353.1439 -352.2696 -348.6324 -341.4887
  -349.4252 -345.5027 -349.4563 -348.4114 -344.0289 -339.5986
  -345.4413 -340.9007 -350.6143 -343.9905 -337.4588 -340.4308
  -349.3837 -347.7803 -339.3221 -352.7309 -349.1086 -338.7972
  -342.6780 -340.7509 -346.5982 -349.3959 -338.5478 -331.5390
  -344.4945 -340.6641 -341.4316 -339.0420 -335.1554 -339.4595
```

Fig. 6.3 BIC values tested

The mean return on the second day is

$$r_{T+2} = 0.00197161 + 0.318231 r_{T+1}(=0.05067) + \epsilon_{T+2}(=0)$$
$$-0.0222004 \epsilon_{T+1}(=0)$$
$$= 0.01810.$$

The predicted mean return quickly approaches to the long-run return of 0.0029.

6.2 GARCH Models

6.2.1 ARCH(q)

Autoregressive Conditional Heteroscedasticity (ARCH) modeling is the predominant statistical technique employed in the analysis of time-varying volatility. In ARCH models, volatility is a deterministic function of historical returns. The original ARCH(q) formulation proposed by Engle [1] models conditional variance as a linear function of the first q past squared innovations:

```
>> Mdl = arima(1,0,1);
EstMdl = estimate(Mdl,r);
[res,~,logL] = infer(EstMdl,r);
stdr = res/sqrt(EstMdl.Variance);

figure
subplot(2,2,1)
plot(stdr)
title('Standardized Residuals')
subplot(2,2,2)
histogram(stdr,10)
title('Standardized Residuals')
subplot(2,2,3)
autocorr(stdr)
subplot(2,2,4)
parcorr(stdr)

    ARIMA(1,0,1) Model:
    --------------------
    Conditional Probability Distribution: Gaussian

                    Standard       t
    Parameter    Value      Error    Statistic
    -----------  ---------  --------- -----------
    Constant   0.00197161  0.00708157    0.278415
     AR{1}      0.318231    0.202951     1.56802
     MA{1}     -0.0222004   0.23931     -0.0927684
    Variance   0.00752877  0.000784057   9.60233
```

Fig. 6.4 Matlab code for ARIMA(1,0,1)

$$\sigma_t^2 = c + \sum_{i=1}^{q} \alpha_i \varepsilon_{t-i}^2.$$

This model allows today's conditional variance to be substantially affected by the (large) squared error term associated with a major market move (in either direction) in any of the previous q periods. It thus captures the conditional heteroscedasticity of financial returns and offers an explanation of the persistence in volatility. A practical difficulty with the ARCH(q) model is that in many of the applications a long length q is needed.

6.2.2 GARCH(p,q)

Bollerslev's Generalized Autogressive Conditional Heteroscedasticity [GARCH (p,q)] specification [2] generalizes the model by allowing the current conditional

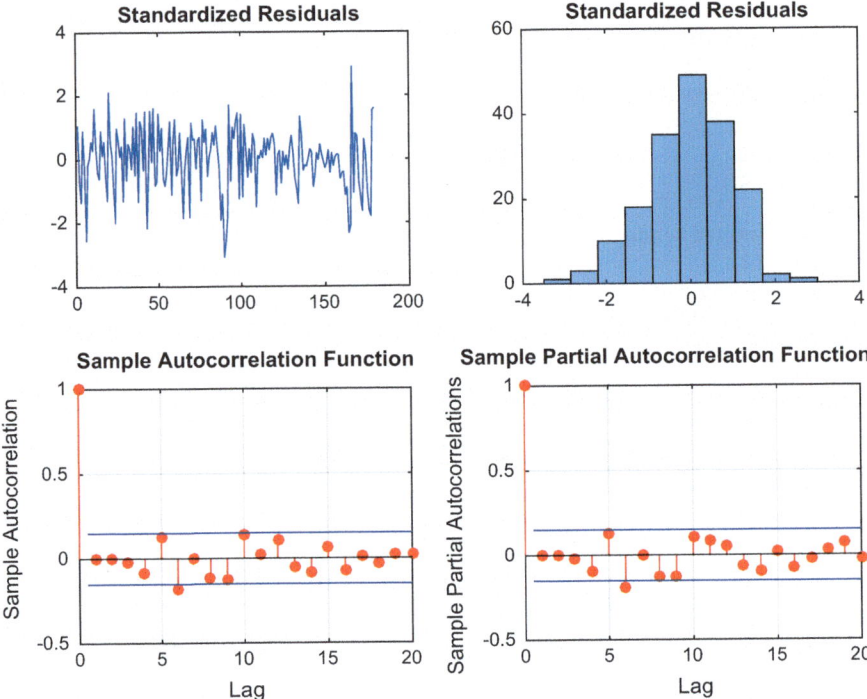

Fig. 6.5 ARIMA(1,0,1) diagnostics output

```
T=length(r);
[E0,V0] = infer(EstMdl,r);
[Y,YMSE,V] = forecast(EstMdl,50,'Y0',r,'E0',E0,'V0',V0);
upper = Y + 1.96*sqrt(YMSE);
lower = Y - 1.96*sqrt(YMSE);

figure
plot(r,'Color',[.75,.75,.75])
hold on
plot(T+1: T+50,Y,'k','LineWidth',2)
plot(T+1:T+50,[upper,lower],'k--','LineWidth',1.5)
xlim([0,T+50])
title('Forecasted Returns')
hold off
```

Fig. 6.6 Matlab code for ARIMA(1,0,1) Forecasts

Fig. 6.7 Graph of forecasts
ARIMA(1,0,1)

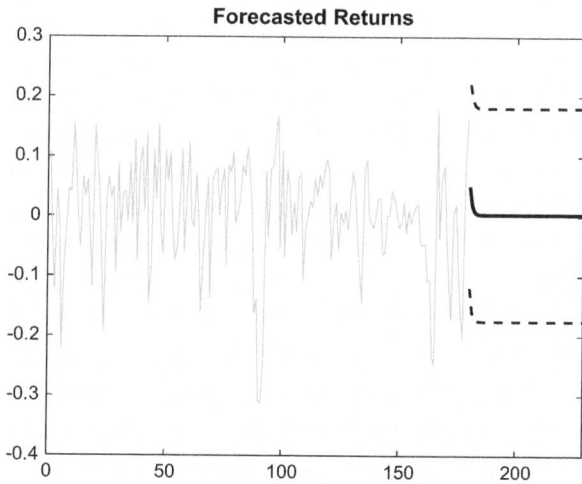

Table 6.1 Numeric forecasts
for the first ten days

Day into future	Proportional change	Dollar forecast
0		48.49
1	0.0507	50.95
2	0.0181	51.87
3	0.0077	52.27
4	0.0044	52.50
5	0.0034	52.68
6	0.0030	52.84
7	0.0029	53.00
8	0.0029	53.15
9	0.0029	53.30
10	0.0029	53.46

variance to depend on the first p past conditional variances as well as the q past
squared innovations. That is:

$$\sigma_t^2 = L + \sum_{i=1}^{p} \beta_i \sigma_{t-i}^2 + \sum_{j=1}^{q} \alpha_j \varepsilon_{t-j}^2,$$

where L denotes the long-run volatility.

By accounting for the information in the lag(s) of the conditional variance in
addition to the lagged t−i terms, the GARCH model reduces the number of
parameters required. In most cases, one lag for each variable is sufficient. The
GARCH(1,1) model is given by:

$$\sigma_t^2 = L + \beta_1 \sigma_{t-1}^2 + \alpha_1 \varepsilon_{t-1}^2.$$

GARCH can successfully capture thick tailed returns and volatility clustering. It can also be modified to allow for several other stylized facts of asset returns.

6.2.3 EGARCH

The Exponential Generalized Autoregressive Conditional Heteroscedasticity (EGARCH) model introduced by Nelson [3] builds in a directional effect of price moves on conditional variance. Large price declines, for instance may have a larger impact on volatility than large price increases. The general EGARCH(p,q) model for the conditional variance of the innovations, with leverage terms and an explicit probability distribution assumption, is:

$$\log \sigma_t^2 = L + \sum_{i=1}^{p} \beta_i \log \sigma_{t-i}^2 + \sum_{j=1}^{q} \alpha_j \left[\frac{|\varepsilon_{t-j}|}{\sigma_{t-j}} - E\left\{ \frac{|\varepsilon_{t-j}|}{\sigma_{t-j}} \right\} \right] + \sum_{j=1}^{q} L_j \left(\frac{\varepsilon_{t-j}}{\sigma_{t-j}} \right)$$

where, $E\{|z_{t-j}|\} E\left\{ \frac{|\varepsilon_{t-j}|}{\sigma_{t-j}} \right\} = \sqrt{\frac{2}{\pi}}$ for the normal distribution, and $E\{|z_{t-j}|\}$ $E\left\{ \frac{|\varepsilon_{t-j}|}{\sigma_{t-j}} \right\} = \sqrt{\frac{v-2}{\pi}} \frac{\Gamma\left(\frac{v-1}{2}\right)}{\Gamma\left(\frac{v}{2}\right)}$ for the Student's t distribution with degree of freedom $v > 2$.

6.2.4 GJR(p,q)

GJR(p,q) model is an extension of an equivalent GARCH(p,q) model with zero leverage terms. Thus, estimation of initial parameter for GJR models should be identical to those of GARCH models. The difference is the additional assumption with all leverage terms being zero:

$$\sigma_t^2 = L + \sum_{i=1}^{p} \beta_i \sigma_{t-i}^2 + \sum_{j=1}^{q} \alpha_j \varepsilon_{t-j}^2 + \sum_{j=1}^{q} L_j S_{t-j} \varepsilon_{t-j}^2$$

where $S_{t-j} = 1$ if $\varepsilon_{t-j} < 0$, $S_{t-j} = 0$ otherwise, with constraints

$$\sum_{i=1}^{p} \beta_i + \sum_{j=1}^{q} \alpha_j + \frac{1}{2} \sum_{j=1}^{q} L_j < 1$$

$$L \geq 0, \beta_i \geq 0, \alpha_j \geq 0, \alpha_j + L_j \geq 0.$$

6.3   Regime Switching Models

Markov regime-switching models have been applied in various fields such as oil and the macroeconomy analysis [4], analysis of business cycles [5] and modeling stock market and asset returns. We consider a dynamic volatility model with regime-switching. Suppose a time series y_t follows an AR (p) model with AR coefficients, together with the mean and variance, depending on the regime indicator s_t:

$$y_t = \mu s_t + \sum_{j=1}^{p} \varphi_j s_t y_t - j + \varepsilon_t, \quad \text{where} \quad \varepsilon_t \sim i.i.d.Normal(0, \ \sigma_{s_t}^2).$$

The corresponding density function for y_t is:

$$f(y_t | s_t, Y_t - 1) = \frac{1}{\sqrt{2\pi\sigma_{s_t}^2}} \cdot \exp\left[-\frac{\omega_t^2}{2\sigma_{s_t}^2} \right] = f(y_t | s_t, y_t - 1, \ldots, y_t - p),$$

where $\omega_t = y_t - \omega s_t - \sum_{j=1}^{p} \varphi_j, \ s_t y_t - j.$

The model can be estimated by use of maximum log likelihood estimation. A more practical approach is to allow the density function of y_t to depend on not only the current value of the regime indicator s_t but also on past values of the regime indicator s_t which means the density function should takes the form of:

$$f(y_t | s_t, S_{t-1}, Y_{t-1}),$$

where $S_{t-1} = \{s_{t-1}, s_{t-2}, \ldots\}$ is the set of all the past information on s_t.

6.3.1   Data

The estimated GARCH for the daily Brent Crude oil data is GARCH(1,1) model as follows:

$$r_t = 0.00521042 + \epsilon_t$$

$$\epsilon_t = z_t \sigma_t$$

$$\sigma_t^2 = 0.00042216 + 0.645574\sigma_{t-1}^2 + 0.354426\epsilon_{t-1}^2$$

The codes in Fig. 6.8 help select a specific model from the GARCH family, i.e., GARCH, EGARCH and GJR-GARCH the best fitted model based on the BIC criteria:

```
>>T = length(r);

logL = zeros(1,3); % Preallocate
numParams = logL;  % Preallocate

Mdl1 = garch('GARCHLags',1,'ARCHLags',1)
[EstMdl1,EstParamCov1,logL(1)] = estimate(Mdl1,r);
numParams(1) = sum(any(EstParamCov1)); % Number of fitted parameters

Mdl2 = egarch('GARCHLags',1,'ARCHLags',1,'LeverageLags',1)
[EstMdl2,EstParamCov2,logL(2)] = estimate(Mdl2,r);
numParams(2) = sum(any(EstParamCov2));

Mdl3 = gjr('GARCHLags',1,'ARCHLags',1,'LeverageLags',1)
[EstMdl3,EstParamCov3,logL(3)] = estimate(Mdl3,r);
numParams(3) = sum(any(EstParamCov3));
[~,bic] = aicbic(logL,numParams,T)
```

Fig. 6.8 Matlab Code to generate GARCH, EGARCH, and GJR models

Table 6.2 GARCH (1,1) conditional variance model

Parameter	Standard value	Error	t statistic
Constant	0.00042216	0.000377294	1.11892
GARCH{1}	0.645574	0.0936578	6.8929
ARCH{1}	0.354426	0.1107	3.20169
Offset	0.00521042	0.0053896	0.966754

Table 6.3 EGARCH(1,1) conditional variance model

Parameter	Standard value	Error	t statistic
Constant	−0.751423	0.490366	−1.53237
GARCH{1}	0.849755	0.098205	8.65287
ARCH{1}	0.553666	0.187644	2.95062
Leverage{1}	−0.0789146	0.0892871	−0.88383
Offset	0.00414701	0.00585085	0.7108786

The resulting models are shown in Tables 6.2, 6.3 and 6.4. All have assumed Gaussian distributions.

Warnings of standard error inaccuracy were given for the GARCH and GJR models, indicating that the only stationary modeling is from EGARCH. With the calibrated EGARCH model, we predict the future 50 days volatility forecast on a daily basis using the codes shown in Fig. 6.9.

Table 6.4 GJR(1,1) conditional variance model

Parameter	Standard value	Error	t statistic
Constant	0.00490018	0.000907454	5.39992
ARCH{1}	0.0406623	0.141255	0.287864
Leverage{1}	0.566574	0.258471	2.19202
Offset	0.0108832	0.00650394	1.67332

```
T=length(r);
V0 = infer(EstMdl2,r);

E0 = (r-EstMdl2.Offset);



V= forecast(EstMdl2,500,'Y0',r);


figure
plot(V0,'Color',[.75,.75,.75])
hold on
plot(T+1:T+500,V,'k','LineWidth',2);
xlim([0,T+500])
```

Fig. 6.9 Matlab code for EGARCH(1,1) Forecasts

Figure 6.10 displays the graph of these forecasts:

The forecasted proportional daily increase for the first ten days are shown in Table 6.5.

Note that EGARCH model takes the form of

$$\log \sigma_t^2 = Constant + GARCH * \log \sigma_{t-1}^2 + ARCH$$
$$* \left[\frac{|\epsilon_{t-1}|}{\sigma_{t-1}} - E\left[\frac{|\epsilon_{t-1}|}{\sigma_{t-1}} \right] \right] + Leverage * \frac{\epsilon_{t-1}}{\sigma_{t-1}}$$

Specifically when the disturbance term is normally distributed,

$$E\left[\frac{|\epsilon_{t-1}|}{\sigma_{t-1}} \right] = \sqrt{\frac{2}{\pi}}.$$

To demonstrate the detailed calculation in the above table, for example, the predicted values of conditional variance on the first two days are calculated respectively as

Fig. 6.10 EGARCH(1,1) forecasts

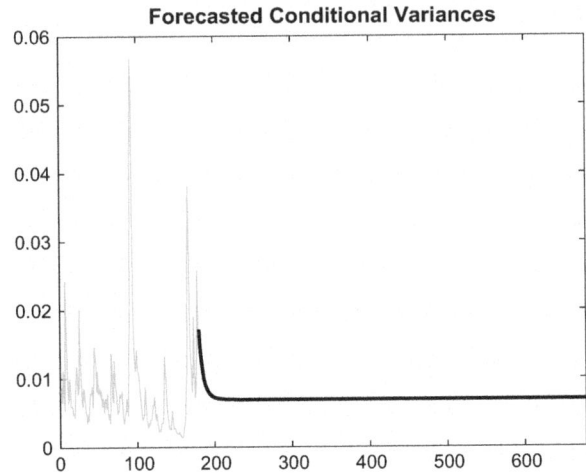

Forecasted Conditional Variances

Table 6.5 EGARCH model forecasts

Day into future	Proportional change	Dollar forecast
0		48.49
1	0.0172	49.32
2	0.0152	50.07
3	0.0136	50.75
4	0.0124	51.38
5	0.0114	51.97
6	0.0106	52.52
7	0.0100	53.05
8	0.0095	53.55
9	0.0091	54.04
10	0.0088	54.51

$$\log \sigma^2_{T+1} = Constant + GARCH * \log \sigma^2_T + ARCH * \left[\frac{|\epsilon_T|}{\sigma_T} - E \left[\frac{|\epsilon_T|}{\sigma_T} \right] \right] + Leverage$$
$$* \frac{\epsilon_T}{\sigma_T}$$

$$\sigma^2_{T+1} = \exp \left(\begin{array}{l} -0.687324 + 0.861882 * \log 0.0172 + 0.540315 * \left(\frac{0.16281}{\sqrt{0.0172}} - \sqrt{\frac{2}{\pi}} \right) \\ + (-0.0922974) * \frac{0.16281}{\sqrt{0.0172}} \end{array} \right) = 0.0172$$

$$\log \sigma^2_{T+2} = Constant + GARCH * \log \sigma^2_{T+1}$$

$$\sigma_{T+2}^2 = \exp(-0.687324 + 0.861882 * \log 0.0172) = 0.0152.$$

Similar to the predicted mean return pattern, the predicted volatility approaches to the long-run volatility quickly as can be seen from the prediction figure.

6.4 Summary

Autoregressive models provide an additional forecasting tool, useful when there are complex patterns in the data. They call for larger data sets, although they can work with as few as 100 observations. This chapter has sought to demonstrate Matlab code enabling the reader to apply both ARIMA and GARCH models, to include parameter diagnosis.

References

1. Engle RF (1982) Autoregressive conditional heteroscedasticity with estimates of variance of United Kingdom inflation. Econometrica 50:987–1008
2. Bollerslev T (1986) Generalized autoregressive conditional heteroskedasticity. Journal of Econometrics 31:307–327
3. Nelson DB (1991) Conditional heteroskedasticity in asset returns: a new approach. Econometrica 59:347–370
4. Raymond JE, Rich RW (1997) Oil and the macroeconomy: a markov state-switching approach. J Money, Credit Banking 29(2):193–213
5. Hamilton JD (1989) A new approach to the economic analysis of nonstationary time series and the business cycle. Econometrica 57(2):357–384

Chapter 7
Classification Tools

Data mining uses a variety of modeling tools for a variety of purposes. Various authors have viewed these purposes along with available tools (see Table 7.1). There are many other specific methods used as well. Table 7.1 seeks to demonstrate that these methods come from both classical statistics as well as from artificial intelligence. Statistical techniques have strong diagnostic tools that can be used for development of confidence intervals on parameter estimates, hypothesis testing, and other things. Artificial intelligence techniques require fewer assumptions about the data, and are generally more automatic, although they might do things that their programmers had not anticipated. The idea of knowledge discovery is a radical change from classical statistical hypothesis testing (although data mining can and does use both), and fits with the idea of data mining big data in our brave new connected world.

Regression comes in a variety of forms, to include ordinary least squares regression, logistic regression (widely used in data mining when outcomes are binary), and discriminant analysis (used when outcomes are categorical and predetermined).

The point of data mining is to have a variety of tools available to assist the analyst and user in better understanding what the data consists of. Each method does something different, and usually this implies a specific problem is best treated with a particular algorithm type. However, sometimes different algorithm types can be used for the same problem. Most involve setting parameters, which can be important in the effectiveness of the method. Further, output needs to be interpreted.

Neural networks have relative disadvantages in dealing with very large numbers of variables, as the computational complexity increases dramatically. Genetic algorithms require specific data structure for genetic algorithms to operate, and it is not always easy to transform data to accommodate this requirement. Another negative feature of neural networks is their hidden nature. Due to the large number of node connections, it is impractical to print out and analyze a large neural network model. This makes it difficult to transport a model built on one system to another system. Therefore, new data must be entered on the system where the neural

© Springer Science+Business Media Singapore 2017
D.L. Olson and D. Wu, *Predictive Data Mining Models*,
Computational Risk Management, DOI 10.1007/978-981-10-2543-3_7

Table 7.1 Data mining modeling tools

Algorithms	Functions	Basis	Task
Cluster detection	Cluster analysis	Statistics	Classification
Regression	Linear regression	Statistics	Prediction
	Logistic regression	Statistics	Classification
	Discriminant analysis	Statistics	Classification
Neural networks	Neural networks	AI	Classification
	Kohonen nets	AI	Cluster
Decision trees	Association rules	AI	Classification
Rule induction	Association rules	AI	Description
Link analysis			Description
	Query tools		Description
	Descriptive statistics	Statistics	Description
	Visualization tools	Statistics	Description

network model was built in order to apply it to new cases. This makes it nearly impossible to apply neural network models outside of the system upon which they are built.

This chapter will look at classification tools, beginning with logistic regression, moving on to support vector machines, neural networks, and decision trees (to include random forests). We will demonstrate with R, which also includes boosting. Classification uses a training data set to identify classes or clusters, which then are used to categorize data. Typical applications include categorizing risk and return characteristics of investments, and credit risk of loan applicants. We will use a bankruptcy data file.

7.1 Bankruptcy Data Set

This data concerns 100 US firms that underwent bankruptcy [1]. All of the sample data are from the US companies. About 400 bankrupt company names were obtained using google.com, and the next step is to find out the Ticker name of each company using Compustat database. This data set includes companies bankrupted during Jan. 2006 and Dec. 2009. This overlaps the 2008 financial crisis, and 99 companies left after that period. After getting the company Ticker code list, the Ticker list was submitted to the Compustat database to get the financial data ratios during Jan. 2005–Dec. 2009. Those financial data and ratios are factors that we can predict the company bankruptcy. The factors we collected are based on the literature, which contain total asset, book value per share, inventories, liabilities, receivables, cost of goods sold, total dividends, earnings before interest and taxes, gross profit (loss), net income (loss), operating income after depreciation, total revenue, sales, dividends per share, and total market value. The LexisNexis database was used to find the company SEC filling after June 2010, which means that companies are still active today, and then we selected 200 companies from the results and got the company CIK code list. The CIK code list was submitted to the Compustat database to obtain financial data and ratios during Jan. 2005–Dec. 2009, the same period for failed companies.

The data set consists of 1321 records with full data over 19 attributes as shown in Table 4.12. The outcome attribute in bankruptcy, which has a value of 1 if the firm went bankrupt by 2011 (697 cases), and a value of 0 if it did not (624 cases). We divided this data set into records prior to 2009 for training (1178 cases—534 no for not bankrupt, 644 yes for bankrupt), holding out the newer 2009–2010 cases for testing (176 cases, 90 no (not bankrupt) and 53 yes (suffered bankruptcy). Note that it is better if data sets are balanced, as most models are more reliable if the data have roughly equal numbers of outcomes. In severe cases, models will predict all cases to be the majority, which could be considered degenerate, and which provides little useful prediction. This data is relatively well balanced. Were it not, means to rectify the situation include replicating the rarer outcome, or if there is a very large data set, even deleting some of the larger set. Care must be taken to not introduce bias during balancing operations, and the synthetic minority over-sampling technique (SMOTE) is incorporated into the R system. Table 7.2 displays variables in our data set, with the outcome variable No. 17.

In the data mining process using R, we need to first link the data. Figure 7.1 shows the variables in the training data file.

Table 7.2 Attributes in bankruptcy Data

No.	Short Name	Long Name
1	Fyear	Data Year—Fiscal
2	At	Assets—Total
3	Bkvlps	Book Value Per Share
4	Invt	Inventories—Total
5	Lt	Liabilities—Total
6	rectr	Receivables—Trade
7	cogs	Cost of Goods Sold
8	dvt	Dividends—Total
9	ebit	Earnings Before Interest and Taxes
10	gp	Gross Profit (Loss)
11	ni	Net Income (Loss)
12	oiadp	Operating Income After Depreciation
13	revt	Revenue—Total
14	dvpsx_f	Dividends per Share—Ex-Date—Fiscal
15	mkvalt	Market Value—Total—Fiscal
16	prch_f	Price High—Annual—Fiscal
17	bankruptcy	bankruptcy (output variable)

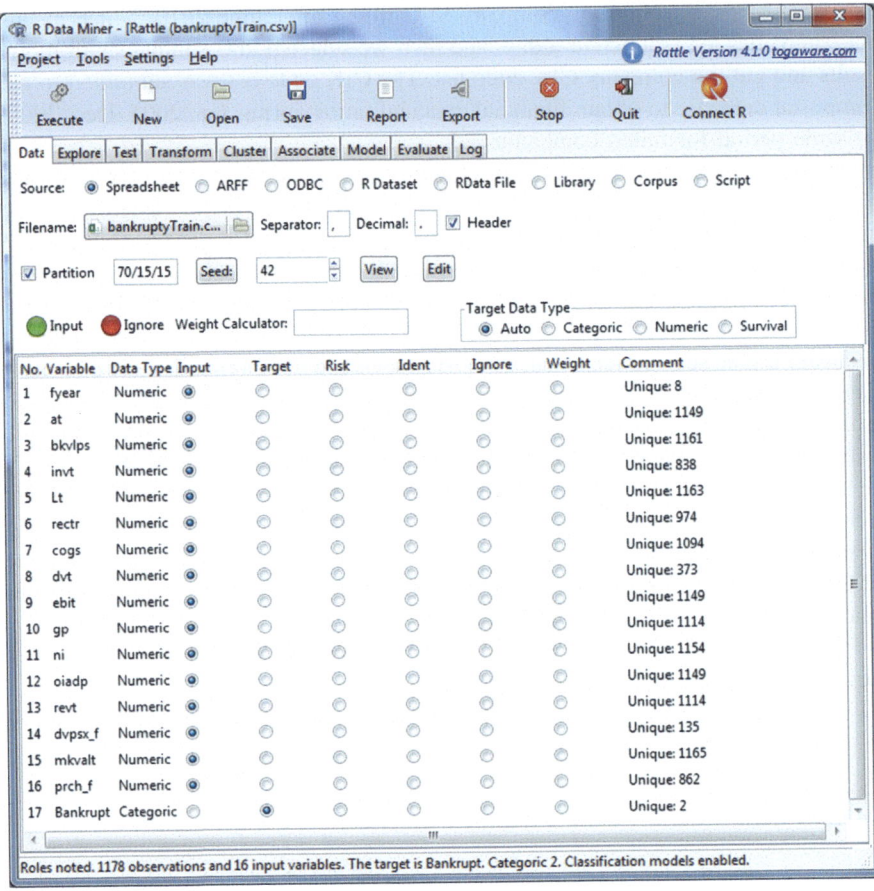

Fig. 7.1 Loading bankruptcyTrain.csv in R

Because variable number 17 is categorical, R presumes (appropriately) that it is the variable to be predicted (target).

7.2 Logistic Regression

In the data set, there are many variables, which probably contain overlapping information. One of the first activities we should apply with regression is to look at correlation, to determine if some variables could be eliminated without losing predictive power. In correlation, we would like high correlation with Bankrupt, but low correlation across candidate independent variables. Table 7.3 shows the correlation matrix obtained in Excel for the training data:

Table 7.3 Correlation matrix—bankruptcyTrain.csv

	fyear	at	bkvlps	invt	Lt	rectr	cogs	dvt	ebit	gp	ni	oiadp	revt	dvpsx_f	mkvalt	prch_f
fyear	1.00															
at	0.06	1.00														
bkvlps	0.05	0.06	1.00													
invt	0.05	0.93	0.09	1.00												
Lt	0.06	1.00	0.06	0.94	1.00											
rectr	0.05	0.99	0.05	0.89	0.99	1.00										
cogs	0.09	0.87	0.01	0.74	0.87	0.87	1.00									
dvt	0.06	0.95	0.04	0.82	0.95	0.98	0.86	1.00								
ebit	0.04	0.74	0.10	0.90	0.75	0.71	0.53	0.67	1.00							
gp	0.06	0.75	0.09	0.88	0.75	0.71	0.60	0.68	0.98	1.00						
ni	−0.04	0.01	0.11	0.15	0.01	0.02	−0.17	0.05	0.44	0.40	1.00					
oiadp	0.04	0.74	0.10	0.90	0.75	0.71	0.53	0.67	1.00	0.98	0.44	1.00				
revt	0.09	0.92	0.04	0.88	0.92	0.90	0.95	0.88	0.77	0.82	0.04	0.77	1.00			
dvpsx_f	−0.01	0.02	0.07	0.02	0.02	0.02	0.02	0.04	0.02	0.02	0.01	0.02	0.02	1.00		
mkvalt	0.07	0.97	0.07	0.87	0.97	0.98	0.87	0.98	0.73	0.75	0.09	0.73	0.91	0.02	1.00	
prch_f	0.12	0.14	0.26	0.18	0.14	0.11	0.13	0.11	0.20	0.22	0.10	0.20	0.18	0.27	0.16	1.00
BankR	−0.05	0.07	−0.01	0.07	0.07	0.06	0.18	0.06	0.07	0.13	−0.02	0.07	0.18	0.03	0.09	0.17

From Table 7.3, we can see that the strongest correlation with BankR in order is with revt, cogs, prch_f, and gp. The others are all below 0.1. The variables revt and cogs have correlation of 0.95, meaning we should only use one of the pair. Thus a safe set of regression variables would be revt, prch_f, and gp. We can deselect the other variables on the Data tab, as shown in Fig. 7.2.

Some data of interest in a regression study may be ordinal or nominal. For instance, in a job application model, sex and college degree would be nominal. In loan application data, the outcome is nominal, while credit rating is ordinal. Since regression analysis requires numerical data, we included them by *coding* the variables. Here, each of these variables is dichotomous; therefore, we can code them as either 0 or 1 (as we did in the regression model for loan applicants). For example, a male is assigned a code of 0, while a female is assigned a code of 1. The

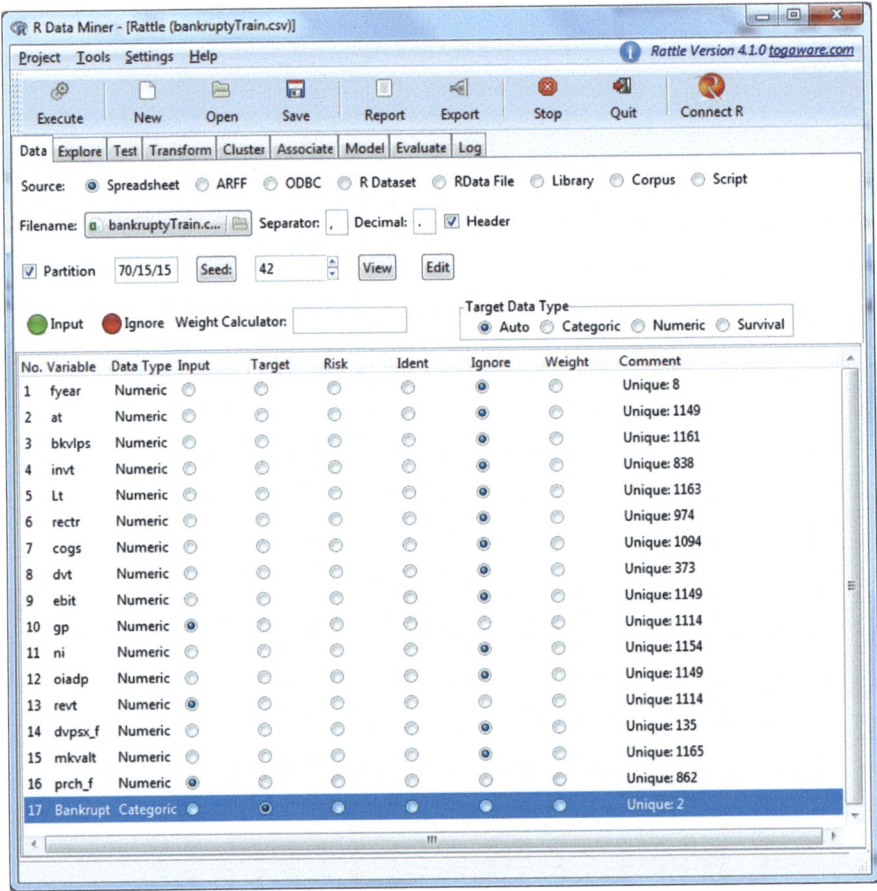

Fig. 7.2 Selecting independent variables

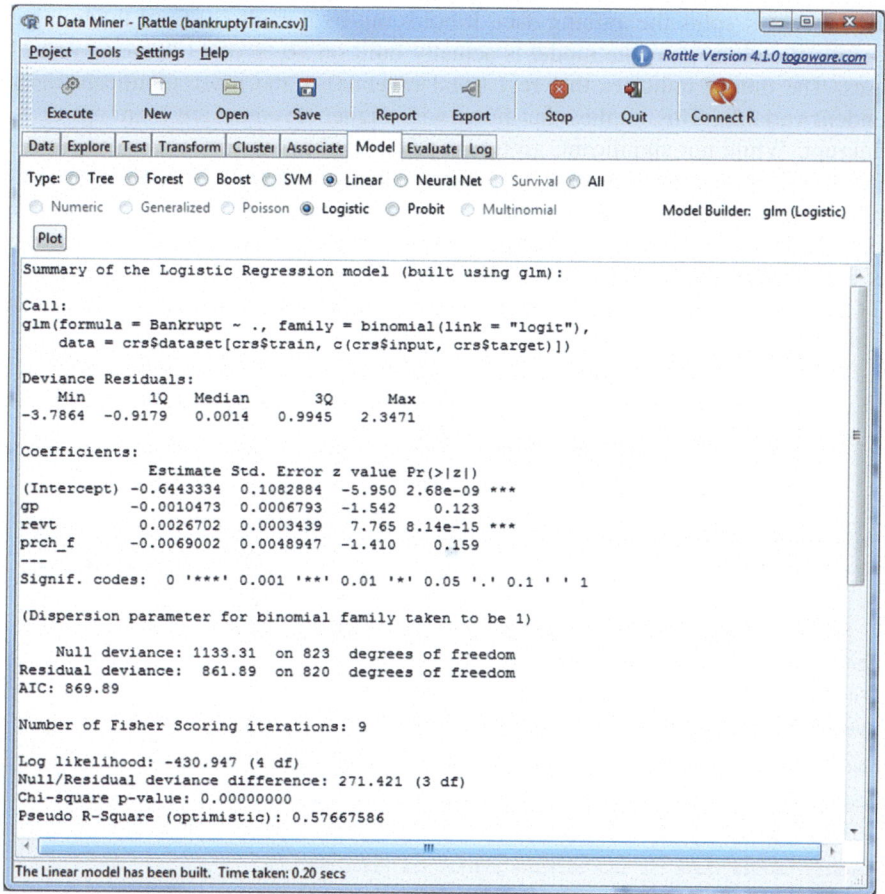

Fig. 7.3 R logistic regression output

employees with a college degree can be assigned a code of 1, and those without a degree a code of 0.

The purpose of logistic regression is to classify cases into the most likely category. Logistic regression provides a set of β parameters for the intercept (or intercepts in the case of ordinal data with more than two categories) and independent variables, which can be applied to a logistic function to estimate the probability of belonging to a specified output class. The formula for probability of acceptance of a case i to a stated class j is:

$$P_j = \frac{1}{1 + e^{\left(-\beta_0 - \sum \beta_i x_i\right)}}$$

where, β coefficients are obtained from logistic regression.

Note that R splits the training data. It holds out by default, 30 % of the training set for validation. Thus the model is actually built on 70 % of 1178 cases, or 824 cases. The output indicates that revt (total revenue) is the most significant independent variable. This implies that firms with higher revenues are more apt to go bankrupt. While not significant, gp (gross profit) makes more intuitive sense—the more gross profit, the less likely the firm will go bankrupt. The third variable, prch_f (high stock price over the year) indicates that the higher the maximum stock price, the less likely the firm would go bankrupt. We can test this model on our test set by selecting the Evaluate tab, as shown in Fig. 7.4.

Clicking on the Execute button yields a coincidence matrix as shown in Table 7.4.

The logistic regression model correctly classified 75 of 90 cases where firms did not go bankrupt (0.833), and 48 of 53 cases where firms did go bankrupt (0.906) for an overall correct classification rate of 0.860.

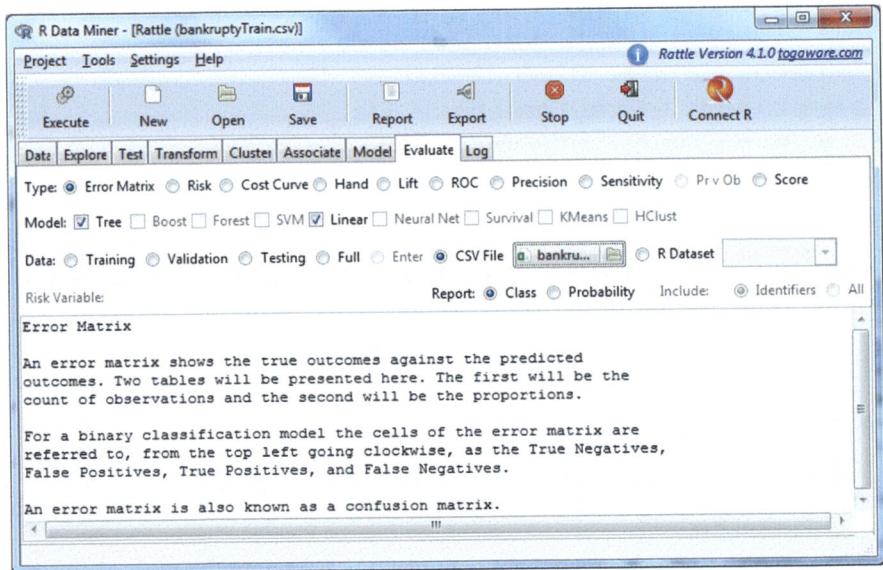

Fig. 7.4 Linking test data set in R

Table 7.4 Coincidence matrix—trimmed logistic regression model		Model no	Model yes	
	Actual no	75	15	90
	Actual yes	5	48	53
		80	63	143

7.3   Support Vector Machines

Support vector machines (SVMs) are supervised learning methods that generate input-output mapping functions from a set of labeled training data. The mapping function can be either a classification function (used to categorize the input data) or a regression function (used to estimation of the desired output). For classification, nonlinear kernel functions are often used to transform the input data (inherently representing highly complex nonlinear relationships) to a high dimensional feature space in which the input data becomes more separable (i.e., linearly separable) compared to the original input space. Then, the maximum-margin hyperplanes are constructed to optimally separate the classes in the training data. Two parallel hyperplanes are constructed on each side of the hyperplane that separates the data by maximizing the distance between the two parallel hyperplanes. An assumption is made that the larger the margin or distance between these parallel hyperplanes the better the generalization error of the classifier will be.

SVMs belong to a family of generalized linear models which achieves a classification or regression decision based on the value of the linear combination of features. They are also said to belong to "kernel methods". In addition to its solid mathematical foundation in statistical learning theory, SVMs have demonstrated highly competitive performance in numerous real-world applications, such as medical diagnosis, bioinformatics, face recognition, image processing and text mining, which has established SVMs as one of the most popular, state-of-the-art tools for knowledge discovery and data mining. Similar to artificial neural networks, SVMs possess the well-known ability of being universal approximators of any multivariate function to any desired degree of accuracy. Therefore, they are of particular interest to modeling highly nonlinear, complex systems and processes.

Generally, many linear classifiers (hyperplanes) are able to separate data into multiple classes. However, only one hyperplane achieves maximum separation. SVMs classify data as a part of a machine-learning process, which "learns" from the historic cases represented as data points. These data points may have more than two dimensions. Ultimately we are interested in whether we can separate data by an $n - 1$ dimensional hyperplane. This may be seen as a typical form of linear classifier. We are interested in finding if we can achieve maximum separation (margin) between the two (or more) classes. By this we mean that we pick the hyperplane so that the distance from the hyperplane to the nearest data point is maximized. Now, if such a hyperplane exists, the hyperplane is clearly of interest and is known as the maximum-margin hyperplane and such a linear classifier is known as a maximum margin classifier.

In R, an SVM model can be selected from the Model tab as shown in Fig. 7.5. Clicking on Execute yields the output shown in Fig. 7.6.

This is not particularly enlightening, but the model can be applied to the test set. Figure 7.6 does indicate that 510 support vectors were generated, with a training error of 16 %. We can test the model on the test data as we did in Fig. 7.4 with the logistic regression model. This yields the coincidence matrix shown in Table 7.5.

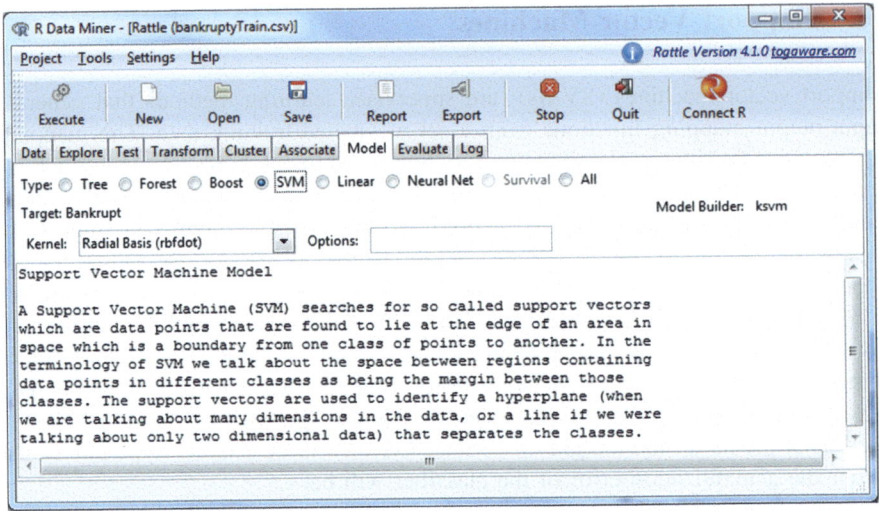

Fig. 7.5 Selecting SVM model in R

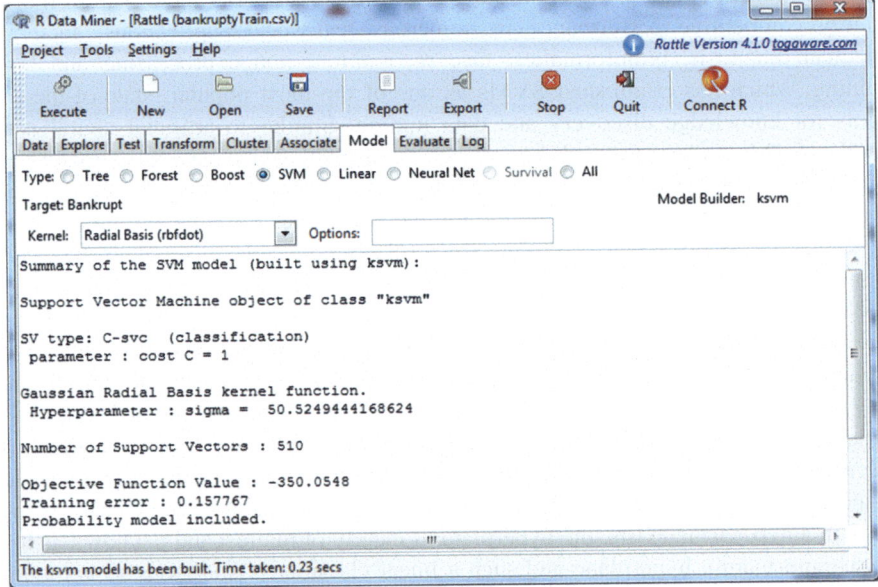

Fig. 7.6 Support vector machine output from R

Table 7.5 Coincidence
matrix for SVM model

	Model no	Model yes	
Actual no	77	13	90
Actual yes	12	41	53
	89	54	143

The SVM model correctly classified 77 of 90 cases where firms did not go bankrupt (0.856), and 41 of 53 cases where firms did go bankrupt (0.774) for an overall correct classification rate of 0.825. It was a bit better at classifying firms that did not go bankrupt relative to the logistic regression model, but worse for those that did.

7.4 Neural Networks

Neural network models are applied to data that can be analyzed by alternative models. The normal data mining process is to try all alternative models and see which works best for a specific type of data over time. But there are some types of data where neural network models usually outperform alternatives, such as regression or decision trees. Neural networks tend to work better when there are complicated relationships in the data, such as high degrees of nonlinearity. Thus they tend to be viable models in problem domains where there are high levels of unpredictability.

Each node is connected by an arc to nodes in the next layer. These arcs have weights, which are multiplied by the value of incoming nodes and summed. The input node values are determined by variable values in the data set. Middle layer node values are the sum of incoming node values multiplied by the arc weights. These middle node values in turn are multiplied by the outgoing arc weights to successor nodes. Neural networks "learn" through feedback loops. For a given input, the output for starting weights is calculated. Output is compared to target values, and the difference between attained and target output is fed beck to the system to adjust the weights on arcs.

This process is repeated until the network correctly classifies the proportion of learning data specified by the user (tolerance level). Ultimately, a set of weights might be encountered that explain the learning (training) data set very well. The better the fit that is specified, the longer the neural network will take to train, although there is really no way to accurately predict how long a specific model will take to learn. The resulting set of weights from a model that satisfies the set tolerance level is retained within the system for application to future data.

The neural network model is a black box. Output is there, but it is too complex to analyze. We ran the model, which generated a model with the 3 inputs, 10 intermediate nodes, and 54 weights which were reported in an output too long to show. Testing this model on the test set obtained the following coincidence matrix shown in Table 7.6.

The neural network model correctly classified 77 of 90 cases where firms did not go bankrupt just as in the SVM model, but 47 of 53 cases where firms did go bankrupt (0.887) for an overall correct classification rate of 0.867. It was thus a better than the SVM model, but the logistic regression did better at predicting bankruptcy.

Table 7.6 Coincidence matrix for neural network model

	Model no	Model yes	
Actual no	77	13	90
Actual yes	6	47	53
	83	60	143

7.5 Decision Trees

Decision trees provide a means to obtain product-specific forecasting models in the form of rules that are easy to implement. These rules have an IF-THEN form, which is easy for users to implement. This data mining approach can be used by groceries in a number of policy decisions, to include ordering inventory replenishment, as well as evaluation of alternative promotion campaigns.

As was the case with regression models and neural networks, decision tree models support the data mining process of modeling. Decision trees in the context of data mining refer to the tree structure of rules (often referred to as association rules). The data mining decision tree process involves collecting those variables that the analyst thinks might bear on the decision at issue, and analyzing these variables for their ability to predict outcome. Decision trees are useful to gain further insight into customer behavior, as well as lead to ways to profitably act on results. The algorithm automatically determines which variables are most important, based on their ability to sort the data into the correct output category. The method has relative advantage over neural network and genetic algorithms in that a reusable set of rules are provided, thus explaining model conclusions. There are many examples where decision trees have been applied to business data mining, including classifying loan applicants, screening potential consumers, and rating job applicants.

Decision trees provide a way to implement rule-based system approaches. All objects in the training set are classified into these branches. If all objects in a branch belong to the same output class, the node is labeled and this branch is terminated. If there are multiple classes on a branch, another attribute is selected as a node, with all possible attribute values branched. An entropy heuristic is used to select the attributes with the highest information. In other data mining tools, other bases for selecting branches are often used.

For the data set with three input variables, we obtain the output shown in Fig. 7.7.

The tree is quite simple—if variable revt (total revenue) was less than 77.696, the decision tree model predicts no bankruptcy (with a confidence of 0.844). Conversely, if total revenue is higher than that limit, the model predicts bankruptcy (with a confidence of 0.860). While it may initially appear counterintuitive that higher revenue leads to higher probability of bankruptcy, this result matches that found in the logistic regression model. In this data set, high volume firms had a

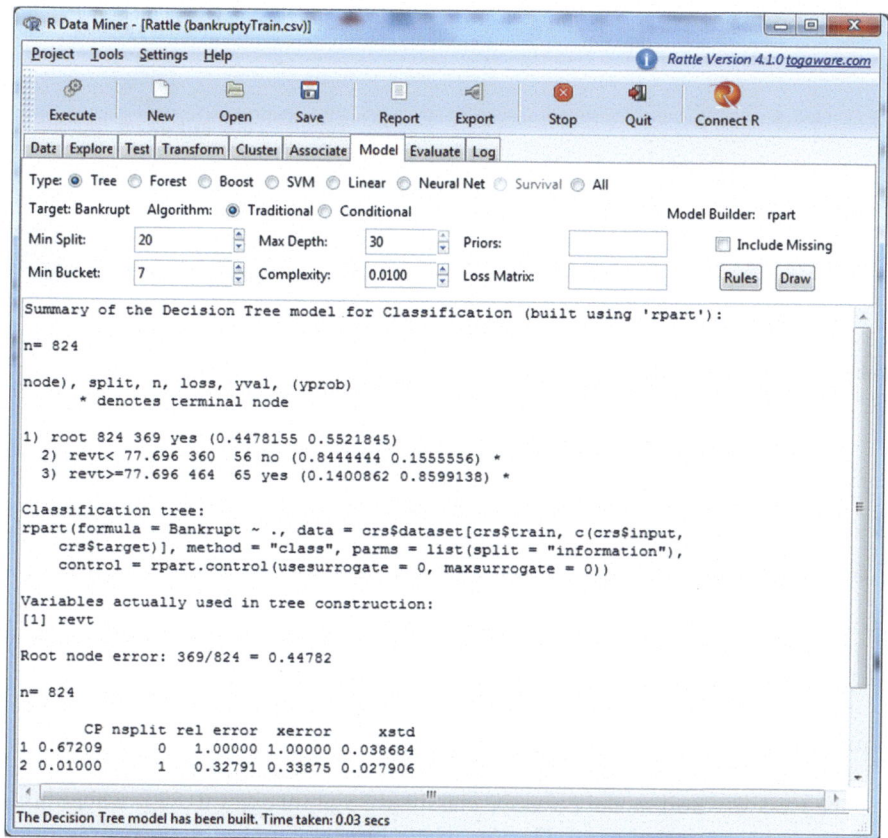

Fig. 7.7 R decision tree output

higher risk of bankruptcy. Decision trees can be simplistic, and we will want to look at the full data set to get a more complete model. But the good feature is that they are blatantly transparent. Any banker or investor can understand this model. Note the Draw button in Fig. 7.7. Clicking on it provides the decision tree exhibit given in Fig. 7.8.

The confidence in no and yes outcomes are shown within the icons in Fig. 7.8, with darker green for greater confidence in a no outcome, and darker yes for higher confidence in a yes (bankruptcy predicted) outcome. Testing this model yields the coincidence matrix given in Table 7.7.

This decision tree model correctly classified 75 of 90 cases (0.833) where firms did not go bankrupt and 48 of 53 cases where firms did go bankrupt (0.906) for an overall correct classification rate of 0.860.

Fig. 7.8 R decision tree

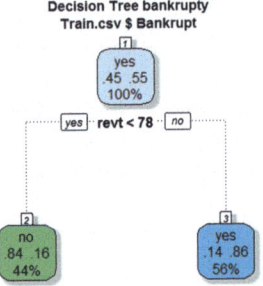

Decision Tree bankrupty
Train.csv $ Bankrupt

Table 7.7 Simple decision
tree coincidence matrix

	Model no	Model yes	
Actual no	75	15	90
Actual yes	5	48	53
	80	63	143

7.6 Random Forests

R provides a description of random forests (see Fig. 7.9), which are essentially
multiple decision trees.

As Fig. 7.9's last paragraph suggests, we can see how the method reduces model
error in Fig. 7.10.

This model gave the coincidence matrix shown in Table 7.8.

The random forest model correctly classified 75 of 90 cases (0.833) where firms
did not go bankrupt and 42 of 53 cases where firms did go bankrupt (0.792) for an
overall correct classification rate of 0.818. Random forests can be expected to do
better than decision trees on training data, but they won't necessarily be better with
test data, as is the case here. This is a case of simpler being better and more reliable.

7.7 Boosting

Boosting is a means to run multiple models with different weights to reduce error,
much in the spirit of neural networks, or as we have just seen, random forests. All
involve intense computation searching for reduced error over the training set.
Figure 7.11 gives R's short description of Adaptive Boosting:

Unfortunately, this approach doesn't always work better on test sets. Table 7.9
gives the coincidence matrix for booting.

The boosted model correctly classified 74 of 90 cases (0.822) where firms did
not go bankrupt and 44 of 53 cases where firms did go bankrupt (0.830) for an
overall correct classification rate of 0.825.

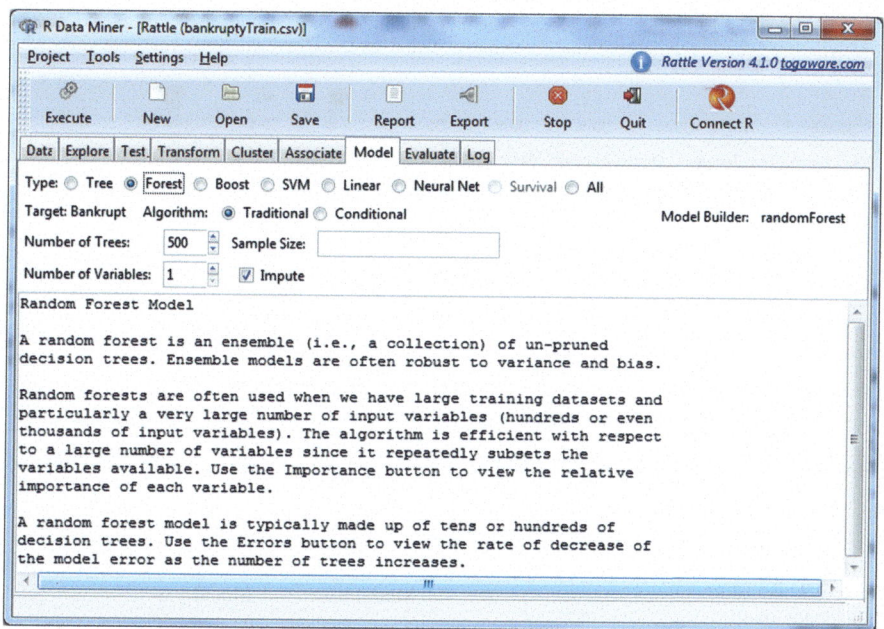

Fig. 7.9 R description of random forests

Fig. 7.10 Random forest error reduction

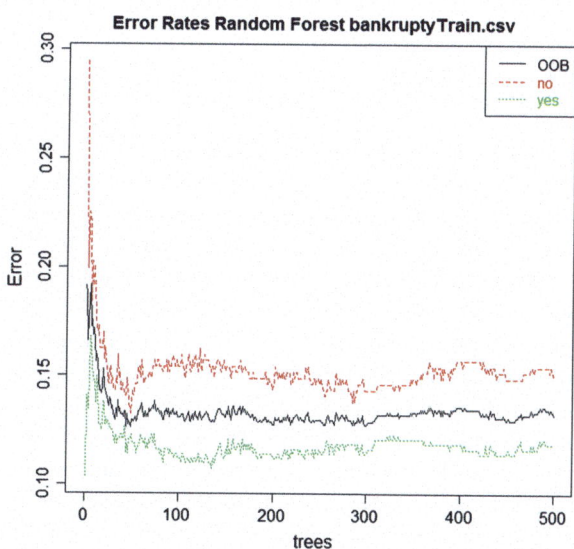

Table 7.8 Random forest coincidence matrix

	Model no	Model yes	
Actual no	75	15	90
Actual yes	11	42	53
	86	57	143

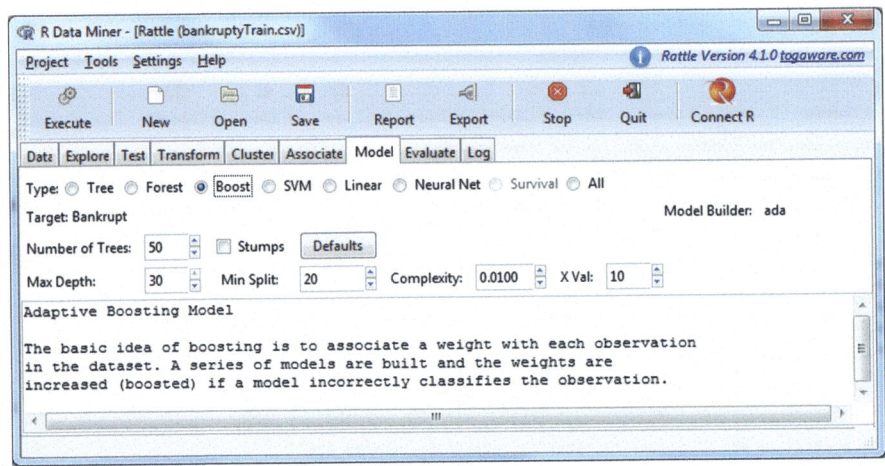

Fig. 7.11 R boosting selection

Table 7.9 Boosting
coincidence matrix

	Model no	Model yes	
Actual no	74	16	90
Actual yes	9	44	53
	83	60	143

7.8   Full Data

We return to the full set of 16 independent variables by going to the Data tab,
marking variables 1 through 16 as Input, and clicking on the Execute tab. We won't
rerun the logistic regression, as we trimmed the number of independent variables to
avoid problems of overlapping information.

Note that two of the variables (including revt which was one of the three we
selected after looking at the correlation matrix) are shown as singular, and thus
excluded from the model. The two most significant variables were rectr (receivables
from trade) and gp (gross profit, seen in the earlier logistic regression output). Both
beta coefficients are positive, indicating that as these variables increase, probability
of bankruptcy increases. For gp, this is a reversal from the model displayed in
Fig. 7.3. It demonstrates the need to carefully look at models, especially as new
data is added. Given the problems of too much overlapping information content in
the logistic regression here, one would probably have more confidence in the model
in Fig. 7.3. Here (Fig. 7.12) the fits quite well, but is potentially unstable due to
overlapping information content among variables. It does well on our test set,
however, as in Table 7.10.

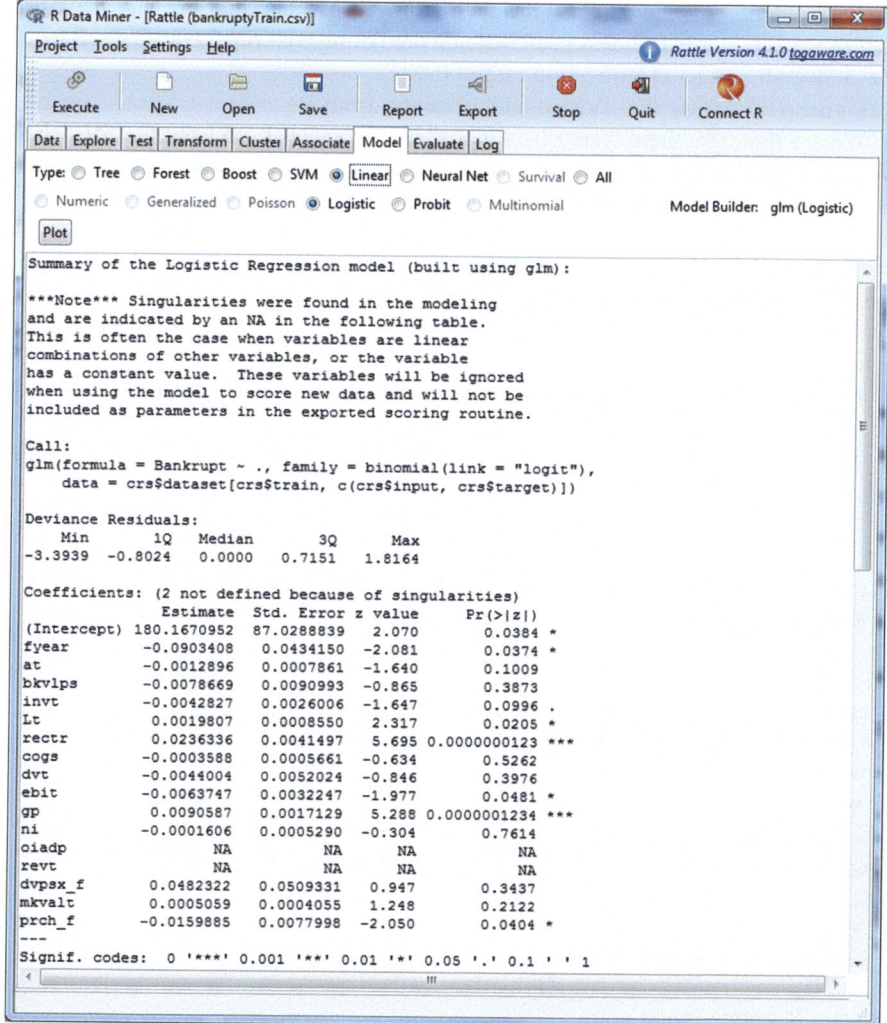

Fig. 7.12 Full logistic regression

Table 7.10 Logistic regression coincidence matrix—all input variables

	Model no	Model yes	
Actual no	79	11	90
Actual yes	4	49	53
	83	60	143

This model was quite good at predicting firms that did not go bankrupt (79/90 = 0.878 correct). It was even better at predicting bankruptcy (49/53 = 0.925), for an overall correct classification rate of 0.902.

Figure 7.13 shows the SVM model obtained from R.

We see that here there were 607 support vectors created, up from the 510 when we only used three input variables. Training error, however, is worse, now at 21 % as opposed to 16 % before. Table 7.11 gives the test results.

The SVM model correctly classified 75 of 90 cases where firms did not go bankrupt, and 31 of 53 cases where firms did go bankrupt. This is slightly worse for those firms that did not go bankrupts, and quite a bit worse for those that did. Overall correct classification dropped from 0.825 to 0.741. More is not always better.

Rerunning the neural network model used 16 inputs with 10 intermediate nodes, generating 197 weights. Table 7.12 shows the results.

The full neural network model correctly classified 75 of 90 cases where firms did not go bankrupt, and 46 of 53 cases where firms did go bankrupt.

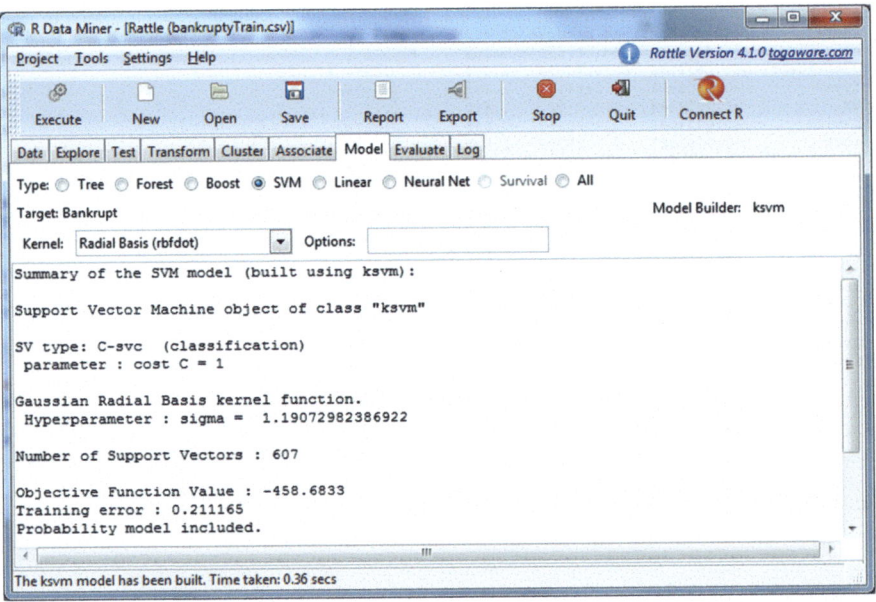

Fig. 7.13 R SVM output for full data

Table 7.11 Coincidence matrix for SVM over all variables

	Model no	Model yes	
Actual no	75	15	90
Actual yes	22	31	53
	97	46	143

Table 7.12 Neural network coincidence matrix—full variables

	Model no	Model yes	
Actual no	75	15	90
Actual yes	7	46	53
	82	61	143

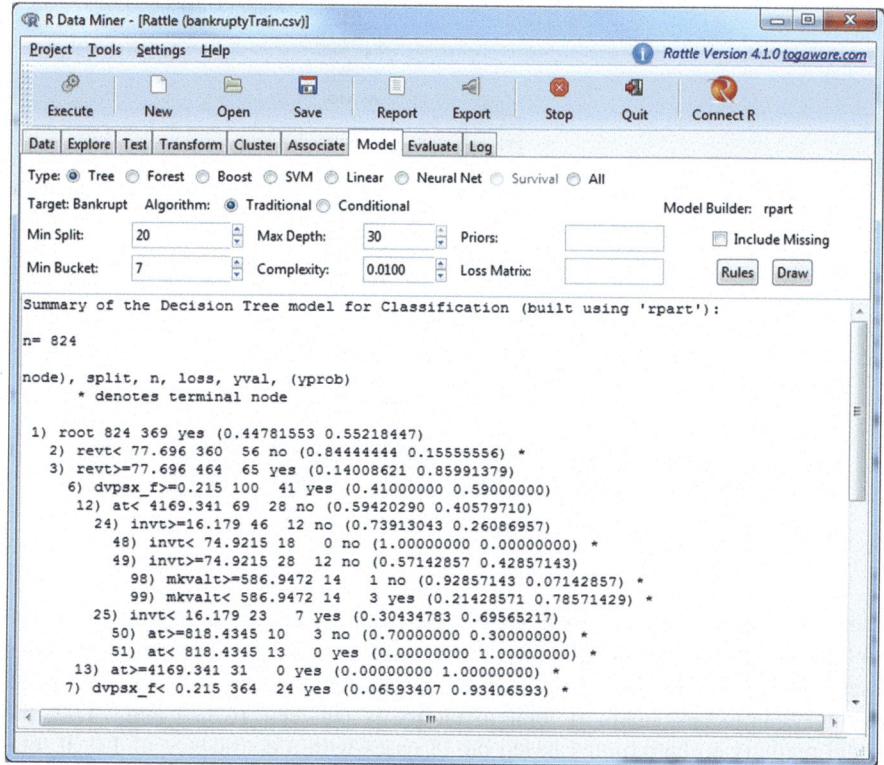

Fig. 7.14 R decision tree results

The decision tree output is shown in Fig. 7.14.

Figure 7.15 displays this tree graphically.

The top rule is essentially identical to that of Fig. 7.7's simpler decision tree. Other variables considered are dvpsx_f (dividends per share). If this value is less than 0.215, the model predicts bankruptcy with a confidence of 0.934. Figure 7.14 shows that if this figure is greater than 0.215, then the tree continues by looking at at (total assets). Note that in Fig. 7.15, the icon is light blue because 59 of 100 cases left on this branch were bankrupt in the training set. Continuing on, if total assets are higher, bankruptcy is predicted with a confidence level of 1.0 (based on only 31 cases). If total assets are less than 4169, variable invt (total inventories) is considered, branching in two directions. If inventories are lower, variable at (total

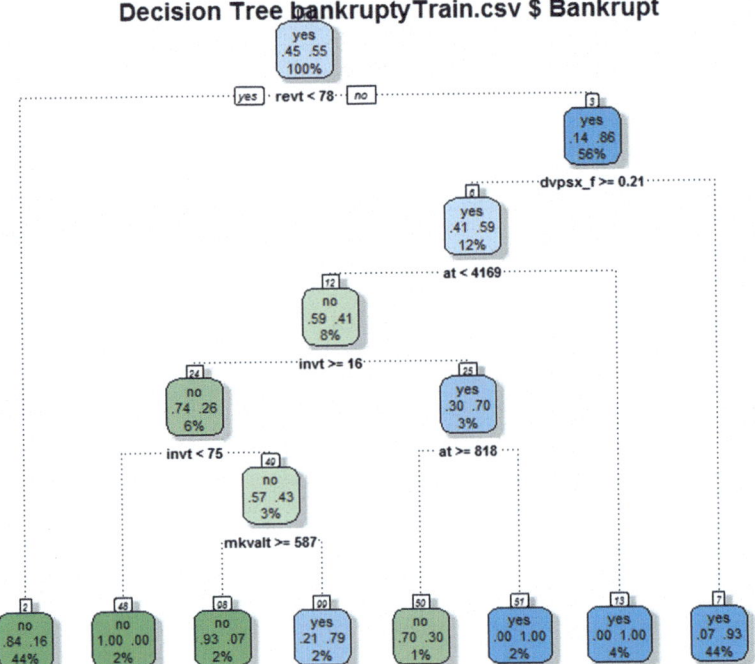

Fig. 7.15 R decision tree

assets) is considered, and if total assets are 818 or greater the model gives a 0.700 confidence of no bankruptcy, as opposed to a 1.0 confidence of bankruptcy if total assets are below 818 (based on only 13 cases). On the other branch, invt (total inventory) branches again. If total inventory is between 16.179 and 74.92, the model predicts no bankruptcy based on 18 cases with a confidence of 1.0. If total inventory is greater than 74.92, variable mkvalt (market value) is considered. Low market value firms on this branch are predicted to go bankrupt with confidence of 0.786 (based on 17 cases), as opposed to a 0.928 confidence of not going bankrupt if market value is higher than the limit. This more complete model, using 5 input variables, yielded the test results shown in Table 7.13.

Table 7.13 Full variable decision tree coincidence matrix		Model no	Model yes	
	Actual no	80	10	90
	Actual yes	7	46	53
		87	56	143

The decision tree model correctly classified 80 of 90 cases (0.889) where firms did not go bankrupt and 46 of 53 cases where firms did go bankrupt (0.868) for an overall correct classification rate of 0.881.

The random forest for this full set of variables is called as shown in Fig. 7.16. Asking for the OOB ROC curve gave the learning displayed in Fig. 7.17.

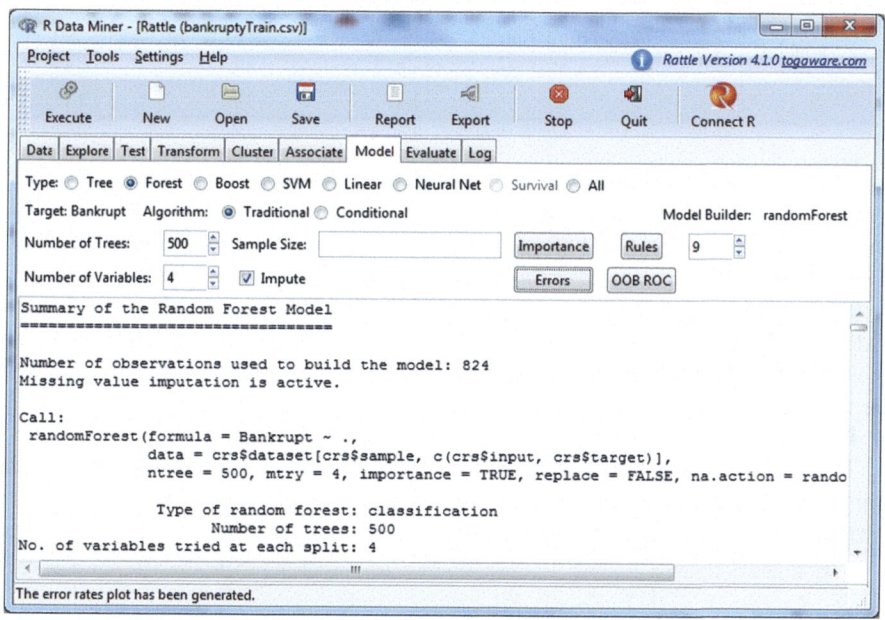

Fig. 7.16 Call for R random forest

Fig. 7.17 Random forest

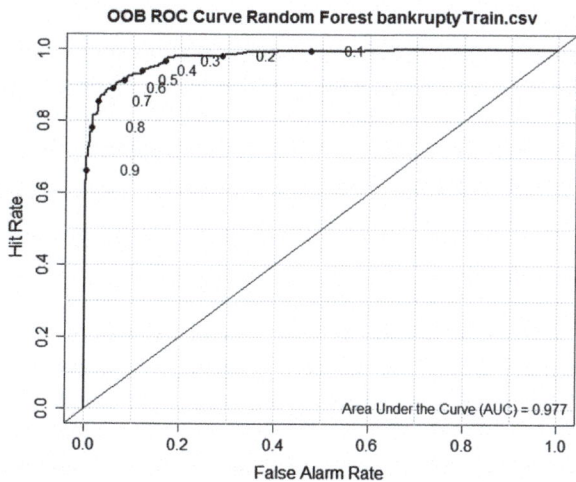

Table 7.14 Coincidence matrix—random forest using all input variables

	Model no	Model yes	
Actual no	79	11	90
Actual yes	4	49	53
	83	60	143

Table 7.15 Coincidence matrix—boosting using all input variables

	Model no	Model yes	
Actual no	79	11	90
Actual yes	5	48	53
	84	59	143

ROC, for historical reasons, stands for "receiver operating characteristics" (from World War II, actually) where it was used to display the area under the curve for true hits versus misses. OOB stands for Out-of-Bag, reflecting a sampling method. Table 7.14 gives test results.

The random forest model correctly classified 79 of 90 cases (0.878) where firms did not go bankrupt and 49 of 53 cases where firms did go bankrupt (0.925) for an overall correct classification rate of 0.895, which is quite good, but again is based on a complex model.

Boosting was also applied using all input variables. Table 7.15 gives test results:

The boosted model correctly classified 79 of 90 cases (0.8278) where firms did not go bankrupt and 48 of 53 cases where firms did go bankrupt (0.906) for an overall correct classification rate of 0.888.

7.9   Comparison

Running all of these models gives a flavor for the R toolbox of classification models. Table 7.16 compares test results for these models for the trimmed data set with 3 input variables (selected based upon correlation).

For the trimmed number of variables, all models yielded very similar results for firms that did not go bankrupt. The SVM and Random Forest models had higher errors in predicting bankruptcy (with boosting error close behind). The neural network model had a slightly higher overall correct classification percentage over logistic regression and decision trees. Use of a classification model, however, also needs to consider stability, as well as ease of implementation. Decision trees and logistic regression have even greater relative advantage on this factor.

Models using all input variables are reported in Table 7.17.

The best overall fit was with the logistic regression using all input variables (which we said was unsound). This model, along with random forests, gave the best predictions of bankruptcy. However, such results should not be taken blindly, and judgment withheld considering expected stability and ease of implementation. More complex models, such as neural networks, random forests, and boosting, will

Table 7.16 Comparison—trimmed variables

Model	Actual no Model no	Actual no Model yes	%	Actual yes Model no	Actual yes Model yes	%	Overall %
Log Reg	75	15	0.833	5	48	0.906	0.860
SVM	77	13	0.856	12	41	0.774	0.825
Neural Net	77	13	0.856	6	47	0.887	0.867
Decision tree	75	15	0.833	5	48	0.906	0.860
Random For	75	15	0.833	11	42	0.792	0.818
Boosting	74	16	0.822	9	44	0.830	0.825

Table 7.17 Comparison—untrimmed variables

Model	Actual no Model no	Actual no Model yes	%	Actual yes Model no	Actual yes Model yes	%	Overall %
Log Reg	79	11	0.878	4	49	0.925	0.902
SVM	75	15	0.833	22	31	0.585	0.741
Neural Net	75	15	0.833	7	46	0.868	0.846
Decision Tree	80	10	0.889	7	46	0.868	0.881
Random For	79	11	0.878	4	49	0.925	0.895
Boosting	79	11	0.878	5	48	0.906	0.888

usually provide better fit on past data. However, part of that is because the more basic models (like logistic regression and decision trees) are subsets of these more complex models. Here the simpler models have the benefit of being easier for users to understand, and we would also contend that they are more stable as new data is encountered. For author preference, go with the decision tree model based on all 16 input variables (it actually only used five inputs). However, the closer review of that model (see Figs. 7.14 and 7.15) demonstrates the instability of models built on small data sets.

Reference

1. Olson DL, Delen D, Meng Y (2012) Comparative analysis of data mining methods for bankruptcy prediction. Decis Support Syst 52:464–473

Chapter 8
Predictive Models and Big Data

In Chap. 1 we tried to use the field of knowledge management as a framework for the connected world of big data that has emerged upon us, and discussed how different kinds of models have contributed to human coping with this highly connected, data-flooded environment. Chapter 2 discussed data sets, which are exploding in the era of big data. One of the first scientific areas to be hit by big data was weather, which continues to be notoriously difficult to predict. But satellites circle the globe and continuously stream data to parallel computers modeling the weather patterns everywhere on earth. While we may not notice it, weather forecasting has improved greatly. The problem is that the weather system on Earth is a highly complex system with severe nonlinearities. Other areas of big data include economic data, to include retail organizations such as Wal-Mart, and application of artificially intelligent real-time trading systems that cause chaos periodically on stock market exchanges. Both involve massive streams of data that are analyzed in real-time, providing Wal-Mart great competitive advantage in retail, and purportedly providing benefit to directors of programmed stock trading. In the medical field, big data is applied in scientific study, as well as in diagnostics by clinical physicians and in administration by medical insurers. Political big data analysis is conducted at a detailed level by both US major political parties, to the extent that they seem to drive candidates to take positions based on real-time popularity polling. Whether this is a good thing or not is a very open question, much like programmed stock trading. But we live in an interconnected world where the possibility of such abuse exists. The first step toward controlling these abuses is understanding the system.

We have examined two broad problem types to which predictive models are applied. Most of the chapters deal with time series forecasting, drawing upon some quite well-established modeling approaches. Time series are very important, where the focus is on a specific variable over time. We used the price of Brent crude oil to demonstrate, but there are many time series of interest to society. For this type of data, linear regression is a starting point, but the problem is that most economic time series are far from linear. We looked at moving average models and

© Springer Science+Business Media Singapore 2017
D.L. Olson and D. Wu, *Predictive Data Mining Models*,
Computational Risk Management, DOI 10.1007/978-981-10-2543-3_8

seasonality as lead-ins to ARIMA modeling. In Chap. 4 we extended examination of regression beyond simple time series to multiple regression. Chapter 5 went a step further and applied decision tree modeling to continuous data forecasting. Chapter 6 applies more advanced modeling techniques to time series data.

Chapter 7 focuses on classification, where the prediction is to take data and attempt to predict which of a finite (usually binary) number of classes into which data is expected to fall. This type of prediction is highly important in many fields. One of the earliest and most effective business applications of data mining is in support of customer segmentation. This insidious application utilizes massive databases (obtained from a variety of sources) to segment the market into categories, which are studied with data mining tools to predict response to particular advertising campaigns. It has proven highly effective. It also represents the probabilistic nature of data mining, in that it is not perfect. The idea is to send catalogs to (or call) a group of target customers with a 5 % probability of purchase rather than waste these expensive marketing resources on customers with a 0.05 % probability of purchase. The same principle has been used in election campaigns by party organizations—give free rides to the voting booth to those in your party; minimize giving free rides to voting booths to those likely to vote for your opponents. Some call this bias. Others call it sound business.

Scalability of models is very important. We have seen that enhanced techniques such as random forests and boosting usually provide improved prediction accuracy. However, this is at the cost of complexity. Furthermore, in the era of big data, scalability is important. Streams of real-time data input can make more complex models difficult to monitor and understand.

For time series forecasting, simple regression works on linear data, but very seldom is real data over time linear. Seasonality is often present and can aid when it is a real factor. Multiple regression would be able to account for more than simple time regression, but as more variables are added, cross-correlation (multicollinearity) becomes a problem. ARIMA and GARCH (assuming computational power to apply high volume) should find more patterns to work with.

In the classification arena, support vector machines would be more complex, and SVM has always been presented as highly useful for sparse data sets, leading us to expect they would not be as good at large scale. Neural networks are not easily interpretable by humans for any set of data, and seem to run on large scale data sets. Simpler models, such as simple linear regression and decision trees should be less affected by large scale data input. Multiple regression is expected to be negatively affected due to the likelihood of cross-correlation across many variables. We would expect decision trees to actually work better in large data sets, because they have more cases upon which to base conclusions. Thus we think that decision trees would be the most scalable classification models.

Data mining offers the opportunity to apply technology to improve many aspects of business. Some standard applications are presented in this chapter. The value of an education is to present you with past applications, so that you can use your imagination to extend these application ideas to new environments.

Data mining has proven valuable in almost every academic discipline. Understanding business application of data mining is necessary to expose business college students to current analytic information technology. Data mining has been instrumental in customer relationship management [1], credit card management [2], banking [3], insurance [4], telecommunications [5], and many other areas of statistical support to business. Business data mining is made possible by the generation of masses of data from computer information systems. Understanding this information generation system and tools available leading to analysis is fundamental for business students in the 21st Century. There are many highly useful applications in practically every field of scientific study. Data mining support is required to make sense of the masses of business data generated by computer technology.

References

1. Ponduri SB, Edara Suma B (2014) Role of information technology in effective implementation of customer relationship management. J Mark Commun 9(3):50–55
2. Snyder N (2014) Mining data to make money. Bank Director 24(3):27–28
3. Bholat D (2015) Big data and central banks. Bank Engl Q Bull 55(1):86–93
4. Donovan K (2015) Mining (and Minding) the data. Best's Review, 46
5. Canzian L, van der Schaar M (2015) Real-time stream mining: online knowledge extraction using classifier networks. IEEE Netw 29(5):10–16

Author Index

© Springer Science+Business Media Singapore 2017   99
D.L. Olson and D. Wu, *Predictive Data Mining Models*,
Computational Risk Management, DOI 10.1007/978-981-10-2543-3

Subject Index

A
Adaptive boosting, 84
AIG, 9
Akaike information criterion (AIC), 58
Amazon, 1
Analysis of variance (ANOVA), 20, 21, 25, 28, 29, 32, 33, 36, 40
Analytics, 3
Association rules, 72, 82
Autoregressive integrated moving average (ARIMA), 7, 12, 13, 15, 55–58, 69, 96

B
Bankruptcy data, 72, 73
Bayesian information criterion (BIC), 56, 58
Big data, 1–3, 9, 10, 95, 96
Boosting, 84, 92, 93, 96
Brent crude oil, 10, 17–19, 23, 36, 95
Business analytics, 2, 5
Business intelligence, 2, 3

C
CART, 45
Classification, 7, 79, 81, 88, 92, 96
Cluster analysis, 72
Coincidence matrix, 78, 79, 82, 85–90
Complexity, 10, 71, 96
Compustat database, 72, 73
Computer support systems, 2
Correlation, 5, 37, 56, 74
Customer relationship management, 4, 97

D
Database management, 1
Data mining, 1, 5, 6, 10, 17, 19, 71, 72, 79, 96, 97
Decision support, 2, 3
Decision tree, 7, 45, 50, 53, 72, 81–84, 90, 92, 93, 96

Descriptive analytics, 5
Descriptive statistics, 5, 6, 72
Diagnostic analytics, 5
Discriminant analysis, 17, 72

E
Enron, 9
Eurostoxx, 10, 14, 35, 37, 45, 46, 50, 52
Executive support, 3
Executive support systems, 2
Exponential generalized autoregressive conditional hetersoscedasticity (EGARCH), 64–66, 68

F
Forecasting, 6, 11, 19, 33, 43, 46, 82, 95, 96

G
Generalized autoregressive conditional heteroskedastic (GARCH), 7, 12, 13, 15, 55, 56, 61, 63, 64, 66, 96
Google, 3, 72

H
Hadoop, 3
Haitian earthquake, 3
Harvard Medical School, 3

I
ID3/C4, 45
Information, 1, 2, 5, 56
Interesting, 5, 11, 12, 15
Internet of Things, 3

K
Knowledge, 1, 4
Knowledge discovery, 5, 79
Knowledge management (KM), 1, 3, 5, 95
Kohonen nets, 72

© Springer Science+Business Media Singapore 2017
D.L. Olson and D. Wu, *Predictive Data Mining Models*,
Computational Risk Management, DOI 10.1007/978-981-10-2543-3

The manufacturer's authorised representative in the EU is Springer
Nature Customer Service Centre GmbH, Europaplatz 3, 69115 Heidelberg,
Germany. If you have any concerns regarding our products, please
contact ProductSafety@springernature.com

Printed and bound by CPI Group (UK) Ltd, Croydon, CR0 4YY
29/04/2026
02099459-0013